Microarray Bioprinting Technology

Moo-Yeal Lee

Microarray Bioprinting Technology

Fundamentals and Practices

Springer

Moo-Yeal Lee
Department of Chemical and Biomedical Engineering
Cleveland State University
Washkewicz College of Engineering
Cleveland, OH, USA

ISBN 978-3-319-46803-7 ISBN 978-3-319-46805-1 (eBook)
DOI 10.1007/978-3-319-46805-1

Library of Congress Control Number: 2016961347

Printed on acid-free paper

This Springer imprint is published by Springer Nature
The registered company is Springer International Publishing AG
The registered company address is: Gewerbestrasse 11, 6330 Cham, Switzerland

Preface

In late 2002, I was exposed to microarray bioprinting technology when my postdoctoral advisor, Professor Jonathan S. Dordick at Rensselaer Polytechnic Institute, received a MicroSys microarray spotter manufactured by Cartesian Technologies (later acquired by Genomic Solutions and DigiLab, Inc.) from VA Medical Center in Albany, NY. Since then, I was fascinated by the potential of microarray bioprinting technology and have focused my research career on developing surface chemistry and methods of printing various biological samples on chip platforms for miniaturized high-throughput toxicology assays. Through several research programs at Rensselaer Polytechnic Institute, Solidus Biosciences, Inc., and Cleveland State University, we have developed highly automatable, high-throughput microarray chip platforms and associated instruments that can be used for miniaturized biochemical- and cell-based assays to assess human metabolism and toxicology. As the lead scientist and the principle investigator, I have been working on the underlying technology for enzyme and cell encapsulation in hydrogel spots and "microarray three-dimensional (3D) bioprinting" on plastic chip platforms for high-throughput, high-content imaging assays. Specific areas of current research include enzyme, virus, and cell printing with inkjet-driven dispensing robots for use in miniaturized 3D cell cultures and organotypic tissue constructs created via layer-by-layer cell printing. In collaboration with engineers from Samsung Electro-Mechanics, Co., a chip platform (S+ micropillar chip and microwell chip) and associated instruments (S+ MicroArrayer and S+ Scanner) have been commercialized. However, implementing microarray bioprinting in miniaturized cell-based assays has been still challenging because of lack of standardized experimental procedures. To this end, I decided to write a book on microarray bioprinting technology with a number of fundamental and practical protocols. Although the majority of protocols provided in this book are based on methodologies we developed and optimized over 10 years, the chapters have all been written with significant contributions from my graduate students and postdoctoral fellow. Without their assistance, it would be impossible for me to complete this book. I hope that this book will serve as a valuable reference manual for graduate students, postdoctoral fellows, scientists, and researchers working in this area of research. We envision that microarray bioprinting technology offers new opportunities for creating highly organized multicellular tissue constructs by precisely dispensing multiple human cell types in hydrogel layers on a chip with printing robots and mimicking the microenvironments of tissues in vivo, thereby potentially revolutionizing regenerative medicine, oncology, and drug discovery.

Cleveland, OH Moo-Yeal Lee

Contents

1 Overview of Microarray Bioprinting Technology 1
Moo-Yeal Lee
1.1 Overview 1
1.2 Microarray Chip Platforms 3
1.2.1 Chips and Instrument 3
1.2.2 Cell Microarrays 6
1.2.3 Enzyme Microarrays 7
1.2.4 Virus Microarrays 11
1.3 Experimental Procedures 12
1.3.1 Cell Spotting 13
1.3.2 Enzyme and Compound Spotting 13
1.3.3 Chip Incubation, Staining, Scanning, and Data Analysis 13
1.4 Advantages and Applications 14
References 16

2 Microarray Spotter and Printing Technologies 19
Akshata Datar, Dong Woo Lee, Sang Youl Jeon, and Moo-Yeal Lee
2.1 Introduction 20
2.1.1 Contact Printing Techniques 20
2.1.2 Non-contact Printing Techniques 21
2.2 Materials 22
2.3 Components of S+ MicroArrayer 22
2.3.1 Main Body Components 23
2.3.2 Externally Connected Components 26
2.4 General Precautions for Operating S+ MicroArrayer 27
2.5 Daily Operations of S+ MicroArrayer 28
2.5.1 Turning 'ON' the System 29
2.5.2 Refilling the Pressure Bottles 29

2.5.3 Washing Tubes, Solenoid Valves, and Ceramic Tips with Ethanol and Water 30
2.5.4 Dispensing Samples on the Micropillar/Microwell Chips with a Work File 32
2.5.5 Replacing Solenoid Valves and Ceramic Tips 34
2.5.6 Turning 'OFF' the System 37
2.6 Detailed Programming for Normal Operation 38
2.6.1 Generating Wash and Dry Sequences 38
2.6.2 Defining Well Plates 38
2.6.3 Registering Chips 41
2.6.4 Registering Spot Layouts 44
2.6.5 Optimizing Dispensing Parameters Using Vision Inspection 47
2.7 Summary 50
References 50

3 Chip Platforms and Chip Surface Treatments 53
Parnian Bigdelou, Alexander Roth, Yana Sichkar, and Moo-Yeal Lee
3.1 Introduction 54
3.1.1 Surface Modification of Micropillar and Microwell Chips ... 54
3.1.2 Surface Modification of Glass Slides 56
3.2 Materials 57
3.3 Protocols 58
3.3.1 Cleaning Microwell Chips with Plasma for Air Bubble Removal 58
3.3.2 Coating Micropillar Chips with Poly(maleic anhydride-*alt*-1-octadecene) (PMA-OD) for Cell Printing 61
3.3.3 Cleaning the Surface of Glass Slides with Acid 61
3.3.4 Coating of Glass Slides with Poly(styrene-*co*-maleic anhydride) (PS-MA) for Cell Printing 63
3.3.5 Coating of Acid-Cleaned Glass Slides with Methyltrimethoxysilane (MTMOS) for Enzyme Printing 64
3.3.6 Coating of Acid-Cleaned Glass Slides with (3-Aminopropyl)trimethoxysilane (APTMS) for Protein Attachment 65
3.3.7 Measurement of Silanization on the Surface of the APTMS Slides with Fluorescein Isothiocyanate (FITC) Labeling 65
3.3.8 Attachment of Glutaraldehyde on the Surface of the APTMS Slides for Protein Attachment 66
3.3.9 Immobilization of Extracellular Matrix (ECM) Proteins on Reactive Surfaces 67
3.4 Summary 68
References 69

4 **Biological Sample Printing** ... 71
Parnian Bigdelou, Alexander Roth, Akshata Datar, and Moo-Yeal Lee
4.1 Introduction ... 71
4.2 Materials ... 73
4.3 Preparation of Stock Solutions ... 74
4.4 Daily Operation of the S+ MicroArrayer for Dispensing Biological Samples ... 76
4.5 Sample Printing Protocols ... 80
4.5.1 Enzyme Printing for Metabolism-Induced Drug Toxicity Assays ... 80
4.5.2 Compound Printing ... 82
4.5.3 Cell Printing ... 85
4.5.3.1 Preparation of Cell Suspension in Growth Medium ... 85
4.5.3.2 Cell Printing for 2D Monolayer Culture on the Micropillar Chip ... 86
4.5.3.3 Cell Printing in Alginate for 3D Cultures ... 87
4.5.3.4 Cell Printing in Matrigel for 3D Cultures ... 90
4.5.3.5 Cell Printing in a Mixture of Alginate and Matrigel for 3D Cultures ... 93
4.5.3.6 Cell Printing in a Mixture of Alginate and Fibrinogen for 3D Cultures ... 95
4.5.3.7 Cell Printing in PuraMatrix for 3D Cultures ... 96
4.5.4 Virus Printing ... 98
4.5.4.1 Measurement of Viral Titer in a 96-Well Plate ... 98
4.5.4.2 Adenoviral Transduction on the Micropillar/ Microwell Chip ... 99
4.6 Inspection of Cells Printed on the Micropillar Chips Using a Bright-Field Microscope ... 101
4.7 Coefficient of Variation (CV) and Z’ Factor for Assay Validation ... 102
4.8 Summary ... 103
References ... 103

5 **High-Content Cell Staining** ... 105
Kyeong-Nam Yu, Pranav Joshi, and Moo-Yeal Lee
5.1 Introduction ... 105
5.1.1 Fluorescent Dyes ... 106
5.1.2 Immunofluorescence (IF) Assays with Antibodies ... 107
5.1.3 Fluorescent Proteins ... 108
5.2 Materials ... 109
5.2.1 Reagents for Fluorescence Staining ... 109
5.2.2 Reagents for Immunofluorescence Staining ... 109
5.2.3 Devices for Cell Staining ... 110
5.2.4 Preparation of Dye Stock Solutions in DMSO ... 111

5.3 Protocols ... 112
5.3.1 Staining Cells with Fluorescent Dyes ... 112
5.3.1.1 Preparation of a Saline Solution ... 112
5.3.1.2 Staining Cells on the Micropillar/Microwell Chip Platform with Fluorescent Dyes ... 112
5.3.2 Staining Cells on the Chip Platform with Fluorophore-Labeled Antibodies ... 116
5.3.2.1 Cell Fixation ... 117
5.3.2.2 Permeabilization of Cell Membranes ... 117
5.3.2.3 Blocking of Nonspecific Binding and Incubation with Primary/Secondary Antibodies for Fluorescence Labeling and Detection ... 118
5.3.3 Measuring the Expression Levels of Drug Metabolizing Enzymes on a Chip ... 119
5.3.3.1 Tyramide Signal Amplification Kit (Life Technologies) ... 121
5.4 Summary ... 122
References ... 122

6 **3D-Cultured Cell Image Acquisition** ... 125
Pranav Joshi, Kyeong-Nam Yu, Emily Serbinowski, and Moo-Yeal Lee
6.1 Introduction ... 125
6.2 Materials ... 127
6.3 Protocols ... 127
6.3.1 Daily Operational Procedures ... 128
6.3.2 Parameter Setting Procedures ... 135
6.3.2.1 Setting the Position of Filters and the Distance of Each Step for Autofocus ... 135
6.3.2.2 Setting XYZ Coordinates for Different Chips and Objective Lenses ... 137
6.4 Summary ... 139
6.5 Appendix ... 139
References ... 141

7 **High-Content Image Analysis** ... 143
Sean Yu, Pranav Joshi, Dong Woo Lee, and Moo-Yeal Lee
7.1 Introduction ... 143
7.2 Materials ... 145
7.2.1 ImageJ ... 145
7.2.2 S+ Chip Analysis ... 145
7.3 Protocols ... 146
7.3.1 3D Cell Image Analysis with ImageJ ... 146
7.3.2 Examples of Image Processing with ImageJ ... 149
7.3.2.1 Hue Filter ... 149
7.3.2.2 Background Subtraction, Brightness Filter, and Region of Interest (ROI) ... 150

7.3.2.3 Outlier Exclusion 151
7.3.2.4 The Performance of the Plugin 151
7.3.3 Image Deconvolution 152
7.3.3.1 Performance of Deconvolution 154
7.3.4 Plotting Dose Response Curves with S+ Chip Analysis 154
7.4 Summary 159
References 159

8 Applications of Microarray Bioprinting 161
Alexander Roth, Emily Serbinowski, and Moo-Yeal Lee
8.1 Introduction 161
8.2 Assay Development for Microarray Bioprinting Technologies 162
8.2.1 Hepatotoxicity Assays 162
8.2.1.1 Phase I and Phase II Drug Metabolizing Enzyme Assays 164
8.2.1.2 Drug Transporter Assays 165
8.2.1.3 Oxidative Stress Assays 166
8.2.2 Neurotoxicity Assays 166
8.2.2.1 Oxidative Stress and Related Assays 167
8.2.2.2 Ion Channel Assays 167
8.2.2.3 Drug Metabolism Assays 168
8.3 Simulation of the *In Vivo* Microenvironment: Liver Applications 169
8.4 Other Applications 171
8.5 Summary 171
References 172

Reviewers

Eben Alsberg Department of Biomedical Engineering and Orthopaedic Surgery, Case Western Reserve University, Cleveland, OH, USA

Jonathan S. Dordick Department of Chemical and Biological Engineering, Rensselaer Polytechnic Institute, 4005 Center for Biotechnology and Interdisciplinary Studies, Troy, NY, USA

Tiago Fernandes Department of Bioengineering, Instituto Superior Técnico (IST), Lisbon, Portugal

Rayton Gerald 3D MicroArray, Inc., Cleveland, OH, USA

Chandrasekhar Kothapalli Department of Chemical & Biomedical Engineering, Cleveland State University, Cleveland, OH, USA

Bosung Ku MBD Korea Co., Ltd., Yongin, South Korea

Seok-Joon Kwon Department of Chemical and Biological Engineering, Rensselaer Polytechnic Institute, 4119 Center for Biotechnology and Interdisciplinary Studies, Troy, NY, USA

Anand K. Ramasubramanian Department of Biomedical Engineering, The University of Texas at San Antonio, San Antonio, TX, USA

Chapter 1
Overview of Microarray Bioprinting Technology

Moo-Yeal Lee

Contents

1.1 Overview 1
1.2 Microarray Chip Platforms 3
1.2.1 Chips and Instrument 3
1.2.2 Cell Microarrays 6
1.2.3 Enzyme Microarrays 7
1.2.4 Virus Microarrays 11
1.3 Experimental Procedures 12
1.3.1 Cell Spotting 13
1.3.2 Enzyme and Compound Spotting 13
1.3.3 Chip Incubation, Staining, Scanning, and Data Analysis 13
1.4 Advantages and Applications 14
References 16

1.1 Overview

Microarray bioprinting refers to printing extremely small amounts of biological samples in patterns on a chip platform, such as functionalized silicone wafers, glass slides, and plastic chips [1, 2]. This technology is mostly based on inkjet printing technologies that allows nanoliter droplet dispensing in high throughput to deposit and pattern biological samples on a chip, including genes, carbohydrates, proteins, viruses, cells, hydrogels, extracellular matrices (ECMs), growth factors, growth

M.-Y. Lee, *Microarray Bioprinting Technology*,
DOI 10.1007/978-3-319-46805-1_1

media, compounds, fluorescent dyes, and other reagents. In general, microarray chips are classified into six different categories depending on biological samples printed, fabricated, or immobilized: (1) gene, (2) carbohydrate, (3) protein, (4) virus, (5) cell, and (6) tissue. These chips have varied applications and methods for fabrication. For example, GeneChip® from Affymatrix (a well-known chip) contains high-density DNA microarrays on silicone wafers manufactured by photolithography technology. Likewise, carbohydrate microarrays contain immobilized oligosaccharides on the surface of glass slides to study interactions between oligosaccharides and antibodies, or other carbohydrate-recognizing proteins and viruses. Z Biotech offers several versions of glycan microarrays on glass slides. Antibody microarrays are a part of protein microarrays, which have a high-density array of antibodies on functionalized glass slides. RayBiotech is one of pioneers who invented various protein microarray-based assays. Our research group invented enzyme microarrays in 2005, which contained 525 spot arrays of drug metabolizing enzymes (DMEs) encapsulated in silicone precursors to simulate metabolism-induced toxicity in human livers [3–7]. The first cell microarrays have been introduced by David M. Sabatini's group in 2001, which is a variant of DNA microarrays and based on reverse transfection of genes on cell monolayers [8]. In 2008, our group demonstrated a method of printing and encapsulating live human cells in 1080 hydrogel spots on a functionalized glass slide for three-dimensional (3D) cell culture and organ-specific toxicity assessment [9–11]. The cell microarray technology has been further improved on a micropillar/microwell chip platform with adenovirus microarrays to control expression levels of DMEs in hepatic cell lines and simulate drug-induced hepatotoxicity [12–17]. Tissue microarrays are somewhat different from other microarrays in terms of the method of fabrication. Typical tissue microarrays are prepared by removing tissue cores in paraffin-embedded tissue samples using a hollow needle, inserting them in a recipient paraffin block in an array pattern, sectioning this block using a microtome, and mounting the slices on glass slides. Tissue microarrays can be a useful tool for immunohistochemistry and fluorescent *in situ* hybridization in cancer research.

Although there have been several books published on microarrays, almost all books focus on DNA and protein microarrays. Thus, there is a need for a practical book focusing on broader microarray bioprinting technologies that cover enzyme, virus, cell, and compound printing. In this book, we provide practical information and standard operating procedures (SOPs) optimized for printing a myriad of biological samples for miniaturized 3D cell cultures and mini-tissue constructs that can be used for biochemical- and cell-based assays to assess human metabolism and toxicology. We illustrate in-depth knowledge in the area of microarray bioprinting for assay miniaturization and provide SOPs for constructing 3D cells on the plastic micropillar/microwell chip via microarray bioprinting. The aims of this practical book are to introduce key fundamentals of microarray bioprinting, chip platforms and associated instruments/devices required, and detailed experimental protocols for chip surface treatment, biological sample printing and incubation, high-content cell staining for mechanistic toxicology, cell image acquisition, and high-throughput data analysis. With several examples of practical applications, we hope that this book can be particularly beneficial to graduate students and researchers working in the areas of tissue engineering, regenerative medicine, disease modeling, high-throughput compound screening, *in vitro* toxicology, and cancer research.

1.2 Microarray Chip Platforms

1.2.1 Chips and Instrument

For robust microarray bioprinting, a highly versatile micropillar/microwell chip system and associated automation instruments have been developed in collaboration with Samsung Electro-Mechanics Co. (SEMCO), South Korea. Therefore, the majority of SOPs developed and optimized are centered on this chip platform for rapidly testing drug candidates and environmental toxicants against human cells isolated from organs. The disposable micropillar/microwell chips with 25 mm × 75 mm in size have been manufactured by plastic injection molding (Fig. 1.1). The new chip platform is a robust and flexible system for 3D cultures of human and animal cells, enzymatic reactions, viral infection, and compound screening. The micropillar chip made of poly(styrene-*co*-maleic anhydride) (PS-MA) or polystyrene coated with poly(maleic anhydride-*alt*-1-octadecene) (PMA-OD) supports 3D cell cultures and comprises an array of human cells for gene expression and toxicity screening (Fig. 1.1A). A single micropillar chip contains 532 micropillars (0.8 mm pillar diameter, 1 mm pillar height, and 1.5 mm pillar-to-pillar distance) onto which an array of 3D human cell cultures (cells entrapped in 60-nL spots of hydrogel) is dispensed using a microarray spotter (Fig. 1.2A). The microwell chip made of polystyrene contains a complementary array of enzymes, recombinant viruses, growth media, reagents, or test compounds (typically 950-nL solutions in the microwells) (Fig. 1.1B). By sandwiching two chips together, various miniaturized biochemical and cell-based

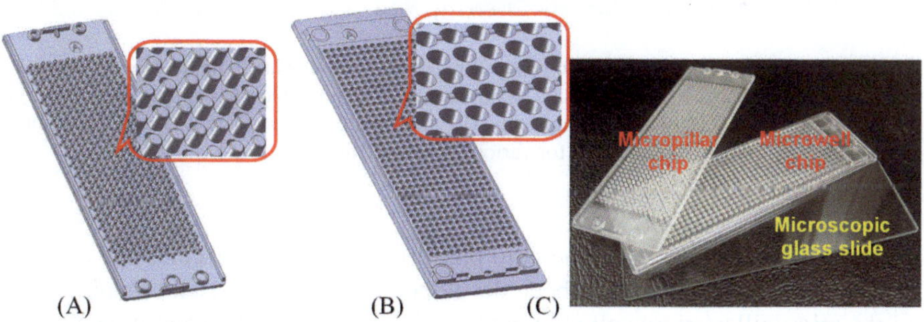

Fig. 1.1 Schematics of (**A**) a micropillar chip and (**B**) a microwell chip. (**C**) Image of micropillar and microwell chips to scale against a microscopic glass slide.

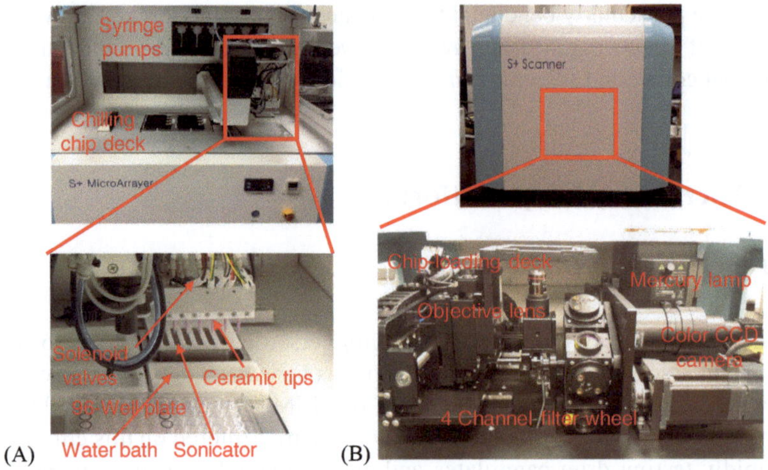

Fig. 1.2 Photographs of (**A**) a microarray spotter (S+ MicroArrayer) with six solenoid valves and ceramic tips for printing biological samples and (**B**) a chip-scanning system (S+ Scanner) for cell image acquisition. The microarray spotter has a printing head with six solenoid valves and a chilling capability that is designed for printing temperature-sensitive hydrogels such as Matrigel. Several human cell types can be spotted in hydrogels such as alginate, fibrinogen, PuraMatrix, and Matrigel onto micropillars (typically 60 nL), and various combinations of biological samples as well as test compounds (typically 950 nL) can be dispensed into microwells. Two chips can be sandwiched together or separated each other consistently and accurately. The sandwiched chips can be incubated in a gas-permeable chamber for cell growth.

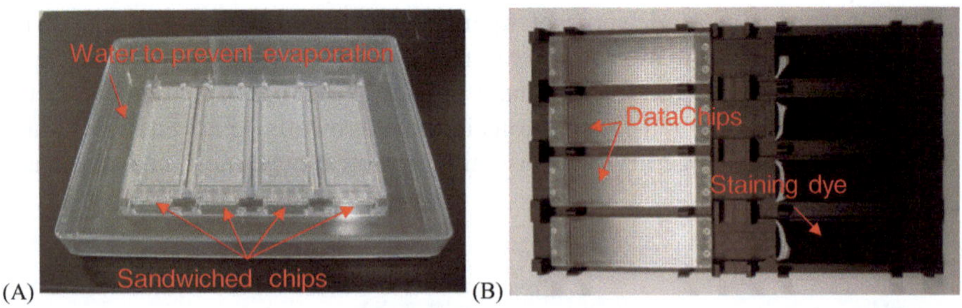

Fig. 1.3 (**A**) Gas-permeable incubation chamber and (**B**) Staining plate. The sandwiched chips are incubated in a chamber that allows good gas exchange. Cell growth occurs during incubation due to uniform gas exchange over the entire chip. The micropillar chips with cells (**DataChip**) can be stained uniformly on a staining plate for various cell-based assays.

assays can be performed on the chip platform (Fig. 1.3). For complete automation, a microarray spotter has been developed to enable consistent manipulation of the micropillar/microwell chip, and a chip-scanning system and image analysis software have been developed to analyze data from the scanned images in a high-throughput fashion (Figs. 1.2 and 1.4).

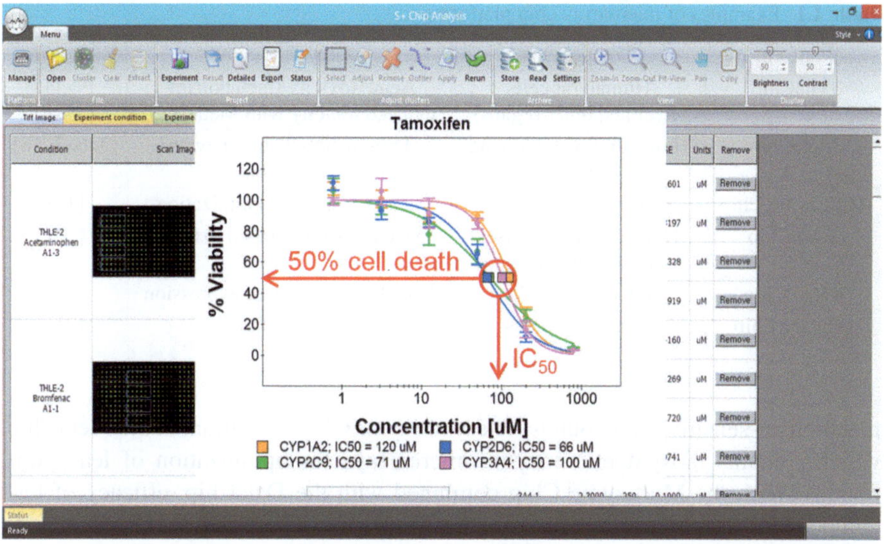

Fig. 1.4 S[+] Chip Analysis from Samsung for cell image analysis. S+ Chip Analysis extracts green and red fluorescent intensities from living and dead cells in each cell spot on the scanned chip, plots sigmoidal dose-response curves with the percentage of live cells against the concentration of the test compound, and then calculates IC_{50} values for each test condition. It takes about 10 min to process three DataChips, which is equivalent to seventy-two individual dose response curves and IC_{50} values.

The micropillar/microwell chip platform represents a promising, high-throughput microscale alternative to conventional *in vitro* multi-well plate platforms and creates new opportunities for rapid and inexpensive assessment of compound efficacy/toxicity at very early phases of drug development. We have been making independent and important contributions to the many interdisciplinary applications of protein-, virus-, and cell-based microarray biochips, including the data analysis toxicology assay chip (**DataChip**) [9–11, 14–16], metabolizing enzyme toxicology assay chip (**MetaChip**) [3, 6, 7], P450 inhibition chip [18], multienzyme lead optimization chip (**Multizyme Chip**) [19], transfected enzyme and metabolism chip (**TeamChip**) [12], RNA interference chip (**RNAi Chip**) [20], and heparin glycan chip (**HepGly Chip**) [21] (Table 1.1). These microarray biochips and instruments will enable low-cost, accurate, high-throughput *in vitro* analysis of drug metabolism and toxicology, and serve as a unique platform for high-throughput toxicology screening in drug discovery.

The chips developed for these microarray technologies can be used independently or in combination with other chip types for various applications. For example, the DataChip represents cell microarrays on a chip that are developed to assess acute toxicity of human organs, whereas the MetaChip represents metabolizing enzyme microarrays on a chip that are specifically designed for analysis of drug metabolism and enzyme inhibition. With the MetaChip combined with the DataChip, metabolism-induced compound toxicity has been demonstrated to simulate hepatotoxicity of

Table 1.1 Examples of microarray chip platforms

Platform	Components	Output/endpoint	References
DataChip	Human/rat cells (liver and other organs)	Acute toxicity (metabolism-induced toxicity with MetaChip)	[9–11, 14–16]
MetaChip	Human/rat metabolizing enzymes	Drug metabolism, Enzyme inhibition	[3, 6, 7, 18]
Multizyme Chip	Biosynthetic enzymes	Lead compound optimization	[19]
TeamChip	Human cells + viruses carrying metabolic genes	Metabolism-mediated toxicity	[12]
RNAi Chip	Interfering RNA	Knock-down of gene expression	[20]
HepGly Chip	Heparin glycans	Anticoagulation	[21]

metabolism-sensitive compounds. The Multizyme Chip is a variant of the MetaChip, which contains biosynthetic enzyme microarrays for optimization of lead compounds. With the Multizyme Chip combined with the DataChip, efficacy of lead compound derivatives generated on the Multizyme Chip can be tested in high throughput. The TeamChip represents recombinant virus microarrays that can be used to express different combinations and levels of DMEs. The RNAi Chip is a variant of the TeamChip developed for knocking down gene expression using short hairpin RNA (shRNA) microarrays. HepGly Chip refer to heparin glycan microarrays developed to optimize the structure of heparin for enhanced anticoagulation. These microarray bioprinting technologies are based on the combination of microscale biocatalysis (for human metabolism of drugs and biosynthesis of lead compounds), human cell culture (for screening major cell types of the body), disease modeling (via over-expression and knock-down of specific genes using adenoviruses, lentiviruses, and shRNAs), materials science (to all microscale fabrication and surface treatment), and automation technology to achieve high-throughput operation with bench top instruments. The most representative cell, enzyme, and virus microarrays will be discussed in more details in the following sections.

1.2.2 Cell Microarrays

Testing efficacy and toxicity of compounds in more predictive cellular models is challenging due to a lack of 3D cell systems that can accommodate high-throughput screening (HTS) assays. The DataChip has the capability to screen drug candidates on 3D-cultured human cells for organ-specific toxicity at speeds commensurate with HTS [9–12]. Human cells in hydrogels such as alginate, fibrinogen, PuraMatrix, and Matrigel are spotted onto the micropillar chip (14 × 38 pillars per chip), and cell spot volumes are as low as 60 nL (Fig. 1.5). Encapsulated cells are grown on individual micropillars of the DataChip when submerged in growth media printed into the microwells of the complementary chip, leading to 3D morphology, which is different from morphology observed in 2D cell culture (Fig. 1.6). Typical cell types accommodated for the DataChip are liver, heart, kidney, bladder, skin, and neural, among others (Table 1.2).

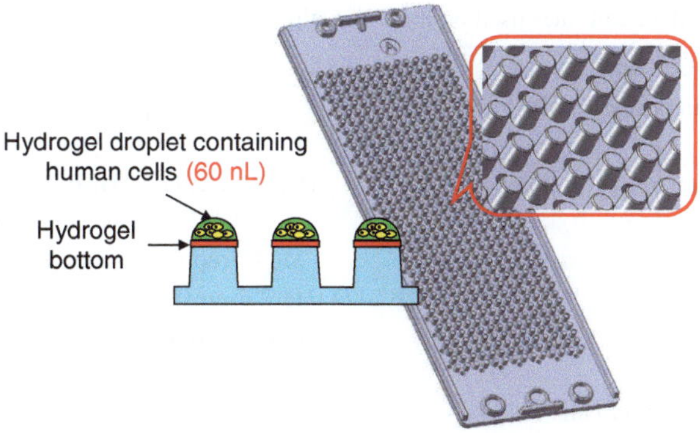

Fig. 1.5 Schematics of the DataChip for organ-specific toxicity.

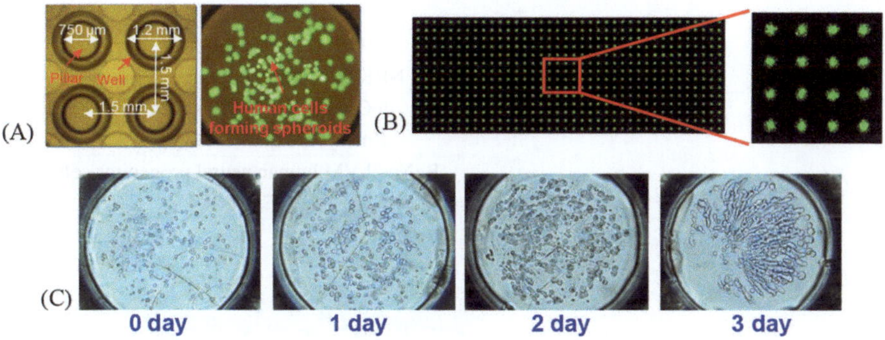

Fig. 1.6 The DataChip—A human 3D cell culture microarray platform: (**A**) Microscopic pictures of the sandwiched chips (*left*) and live Hep3B human hepatoma cell line encapsulated on the micropillar chip (*right*), (**B**) Scanned image of the entire micropillar chip with live Hep3B cells stained with a green fluorescent dye, (**C**) Culture of Hep3B cells on the microwell chip over time, forming 3D spheroids.

1.2.3 Enzyme Microarrays

The MetaChip with active human (or rat) metabolizing enzymes (mixtures or human liver microsomes) encapsulated in a hydrogel matrix has been developed to investigate drug metabolism and enzyme inhibition by compounds (Fig. 1.7). Typical DMEs printed in the MetaChip include cytochrome P450 (CYP450), UDP-glycosyltransferase (UGT), sulfotransferase (SULT), and glutathione S-transferase (GST), among others (Table 1.3). The MetaChip allows the production and screening of human (or animal) metabolites on the chip platform because enzymatic kinetic parameters are comparable with conventional approaches. Since it relies on nanoliter-scale enzyme printing, the approach can save the expense of reagents and test compounds significantly (ca. 100-fold reduction in assay volume compared to the 96-well plate counterpart).

Table 1.2 Breadth of cell lines used on the DataChip

Cell type	Name
Human primary cells	Hepatocytes
	Astrocytes
Human transformed cell lines	Hep3B, HepG2, THLE-2 (hepatoma)
	A293 (adenocarcinoma)
	MCF7 (epithelial adenocarcinoma)
	BT474 (Her2 overexpressing breast adenocarcinoma)
	Mia PaCa-2 (pancreatic adenocarcinoma)
	MCF10A (non-tumorigenic epithelial)
Rat primary cells	Hepatocytes
	Astrocytes
	Cardiomyocytes
Rat transformed cell lines	H411E (hepatoma)
	DITNC1 (astrocytoma)
	H9c2(2–1) (cardiomyocyte)
	NRK-52E (renal proximal tubular)
Others	46C mouse embryonic stem cell
	Embryonic rat neural stem cell
	ReNcell VM human neural progenitor cell
	E. coli

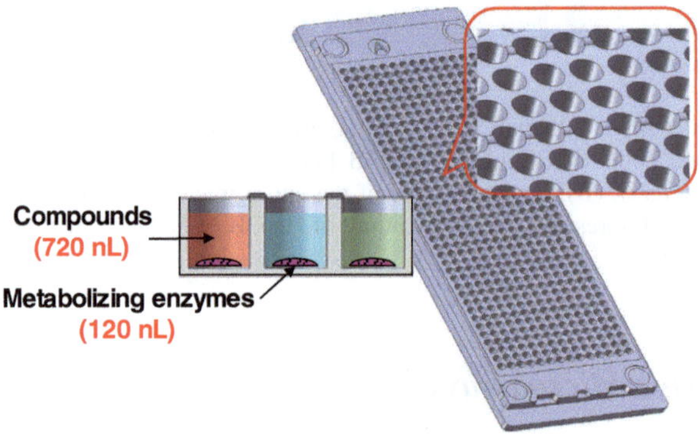

Fig. 1.7 Schematics of the MetaChip for drug metabolism and inhibition.

By printing test compounds and fluorogenic substrates such as BOMCC and EOMCC from Life Technologies, it is possible to assess P450 inhibition by compounds on the MetaChip [18]. The microwell chip can accommodate the measurement of fluorescent dyes at different concentrations (Fig. 1.8). To demonstrate P450 inhibition on CYP450 microarrays, CYP2D6, CYP2C9, and CYP3A4 were printed and immobilized on three different regions of a functionalized glass slide in 21 × 9 arrays. CYP450 isoform-specific inhibitors were printed interspersed with control reaction spots on each region, and the extent of reaction was assessed [18]. Combined

Table 1.3 Breadth of drug metabolizing enzymes (DMEs) used on the MetaChip

Drug metabolizing enzymes	Name
Human Phase I enzymes	Cytochrome P450 (CYP1A2, 2C9, 2C19, 2D6, 2E1, 3A4, 3A5)
	Flavin monooxygenase (FMO1, FMO3, FMO5)
	Monoamine oxidase (MAO-A, MAO-B)
	Others (Myeloperoxidase, Esterase, Epoxide hydrolase, DT diaphorase, NQQ1)
Human Phase II enzymes	UDP-glycosyltransferase (UGT1A1, 1A4, 2B4, 2B7)
	Sulfotransferase (SULT1A1, 1A3, 1B1)
	Glutathione S-transferase (GST)
	N-Acetyltransferase (NAT1 and NAT2)
Pooled human liver enzymes	Human liver microsomes (HLM)
	Human s9 fractions (Hs9)
Rat Phase I enzymes	Suite of rat CYP450s
Pooled rat liver enzymes	Rat liver microsomes (RLM)
	Rat s9 fractions (Rs9)

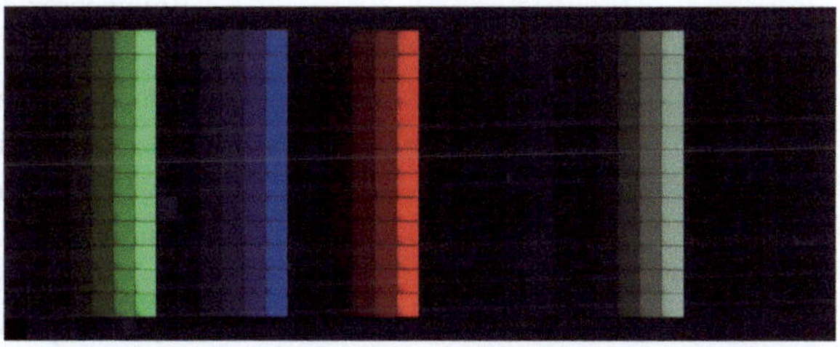

Fig. 1.8 Scanned image of the microwell chip with varying concentrations of fluorescent dyes obtained by Cellomics ToxInsight *In Vitro* Toxicity (IVT) platform.

with the DataChip, the MetaChip has been used for analyzing metabolism-induced toxicity. For example, the DataChip with Hep3B cells was stamped onto the MetaChip containing metabolic enzymes (no enzyme, P450 Mix, All Mix, and HLMs) and compounds (six compounds at six different dosages per compound) [3, 7, 9]. No enzyme contains baculosome® negative control, P450 Mix is a mixture of major P450 isoforms, All Mix is a mixture of P450 Mix and Phase II enzymes, and HLMs are pooled human liver microsomes. This approach allows one to measure not only the toxicity of parent compounds but also the augmented toxicity by drug metabolites, which are critically important to predict human drug metabolism in the livers after drug administration. Thus, it enables early detection of hidden toxicity of metabolites during lead generation, lead optimization, and preclinical phases of drug development.

The Multizyme Chip, the variant of the MetaChip, encompasses microarrays of biosynthetic enzymes on a chip to construct artificial synthetic pathways (Fig. 1.9). Biosynthetically important enzymes are encapsulated into high-density microarray spots that have the capacity to perform large numbers of enzymatic reactions on a single chip, ultimately generating a large number of derivatives starting from a lead compound. Thus, the Multizyme Chip is specifically designed

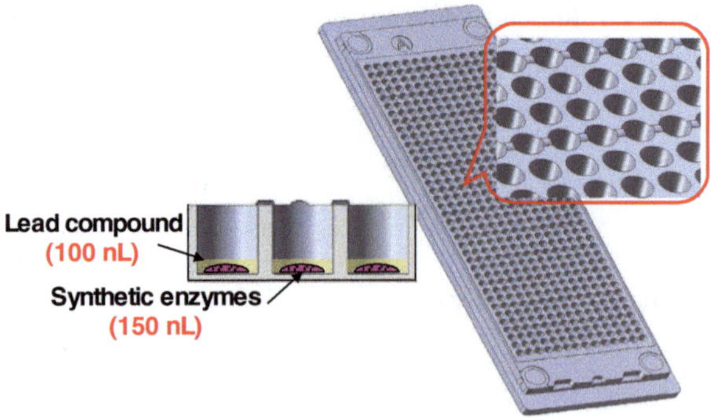

Fig. 1.9 Schematics of the Multizyme Chip for lead compound optimization.

Table 1.4 Breadth of biosynthetic enzymes used on the Multizyme Chip

Enzyme	Name
Peroxidases	Lignin peroxidase
	Soybean peroxidase
	Chloroperoxidase
	Bromoperoxidase
Oxidases	Tyrosinase
	Alcohol oxidase
	Aryl alcohol oxidase
	Cholesterol oxidase
	Bilirubin oxidase
Dehydrogenases	Alcohol dehydrogenase
	Aldehyde dehydrogenase
Transaminases	Transaminase
	Amine transaminase
Others	Oxynitrilase
	Ketoreductase
Type III polyketide synthases (PKSs)	Tetrahydroxynaphthalene synthase (THNS)
	Chalcone synthase (CHS)
Kinases	Tyrosine kinase

for high-throughput derivatization of lead compounds. It is well suited to transform complex lead compounds (e.g., natural products and multifunctional synthetics), as such molecules are difficult to modify selectively and controllably. In combination with the DataChip, it can be used for cell-based efficacy and toxicity screens for the rapid identification of biologically active derivatives. This approach represents a new paradigm for lead optimization by tapping into nature's vast biocatalytic repertoire. A list of biosynthetic enzymes tested on the Multizyme Chip is presented in Table 1.4. The strategy of generating diverse polyketide analogs has been successfully demonstrated using the *in vitro* metabolic pathway microarray consisting of type III polyketide synthases (PKSs) and post-PKS tailoring enzymes [19, 22]. After synthesizing diverse polyketide analogs on the enzyme microarrays, the efficacy of polyketide analog library was assessed by tyrosine kinase assays in high throughput.

1.2.4 Virus Microarrays

The TeamChip encompasses microarrays of recombinant adenoviruses carrying genes for DMEs, which is prepared by printing several recombinant adenoviruses in a combinatorial fashion (Figs. 1.10 and 1.11). By sandwiching the DataChip with the TeamChip, human cells on the micropillar chip can be infected by adenoviruses in the microwell chip, leading to creating human cell microarrays expressing different

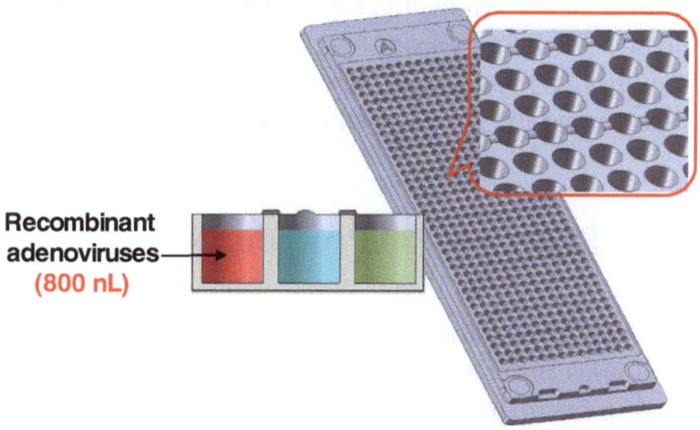

Fig. 1.10 Schematics of the TeamChip for viral gene transduction with the DataChip.

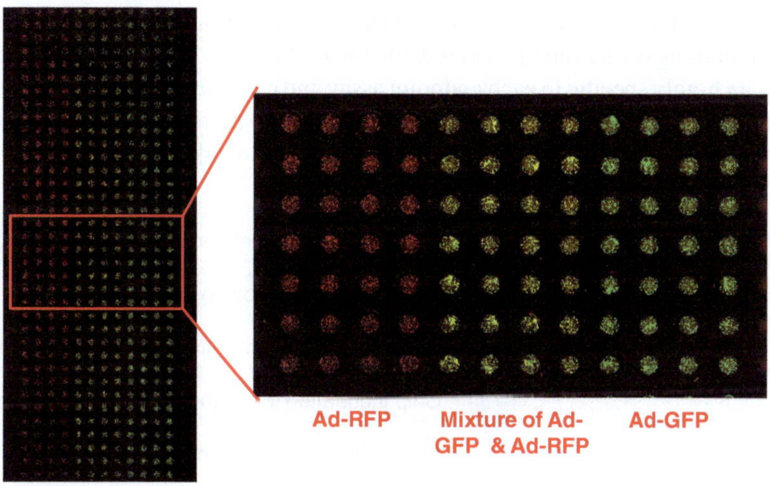

Fig. 1.11 The scanned image of the DataChip containing THLE-2 cells after sandwiching with the TeamChip containing recombinant adenoviruses carrying GFP/RFP genes (Ad-GFP/RFP). Human cells (e.g., THLE-2, Hep3B, HepG2, and Beas-2b) in Matrigel spots were 100 % infected at 60 multiplicity of infection (MOI).

combinations of metabolizing enzymes on the chip [12]. The TeamChip is being developed to mimic the first-pass metabolism of the human liver and to predict enzyme-specific hepatotoxicity, resulting from exposures to various chemicals, environmental pollutants, biologics, and therapeutic molecules or drugs. Thus, the reactivity of target compounds with individual human metabolizing enzymes or combinations of enzymes in the human liver can be assessed and quantified at speeds commensurate with early-stage predictive human toxicity assessment. The TeamChip provides metabolizing enzymes to virtually any human cell type, thereby providing predictions on xenobiotic toxicity in essentially any cell type in the body. By adjusting the expression levels of the various metabolizing enzymes in human cells to match the enzyme levels representative of a subgroup unique to an individual, the TeamChip chip could be tailored for analysis of drug compounds being developed for specific subgroups of the population and for individual patients. Similar to the TeamChip, high-throughput shRNA transfection on the DataChip has been demonstrated using a reverse transfection approach [20]. The DataChip containing CHO and 3T3 cells was transfected with toxic shRNAs in a high-throughput manner and the percent cell death was measured as a proof of concept.

1.3 Experimental Procedures

The highly versatile microarray biochip platform is based on micropillar/microwell structures made by plastic injection molding, which is a more robust and flexible system for mammalian cell culture in 3D, enzymatic reactions, viral infection, and compound screening compared to its counterpart, microtiter plates. The chip platform doesn't require constant mixing or multiple wash steps during a cell-based assay, and solution exchanges can be done by simply transferring the micropillar chip containing cells from one microwell chip to the other. The experimental procedures are highly specific to each endpoint assay and can be modified to meet requirement of the assay [6]. Here we provide a simplified experimental procedure for the DataChip/MetaChip for metabolism-induced toxicity assays (Fig. 1.12).

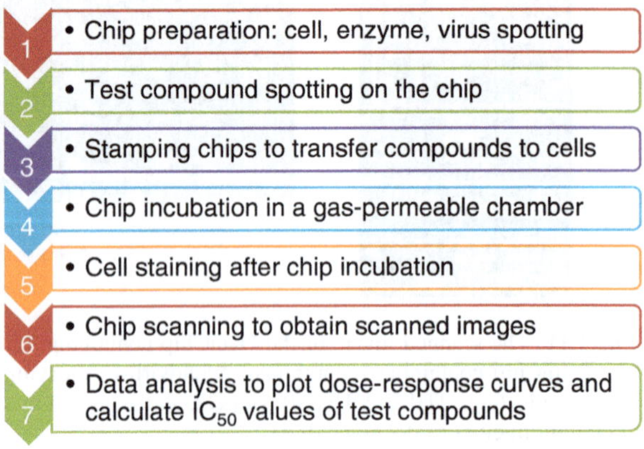

Fig. 1.12 Experimental procedures with the DataChip/MetaChip for toxicology screening,

1.3.1 Cell Spotting

To attach cell spots on the micropillar chip, a mixture of poly-L-lysine (PLL) and barium chloride ($BaCl_2$) is prepared by mixing a 1:2 volume ratio of 0.01 % (wt/vol) PLL and 24 mM $BaCl_2$. The DataChip is prepared by spotting 60 nL of the PLL/$BaCl_2$ mixture onto 532 micropillars followed by printing 60 nL of hepatic cell suspension in alginate on top of the dried PLL/$BaCl_2$ spots. While printing cells, the micropillar chip is placed on a chilling chip deck at 4 °C to retard evaporation of water in the spots. A suspension of cells in low-viscosity alginate is prepared by mixing cell suspension in FBS-supplemented growth medium with alginate solution in distilled water so that the final concentration of cells and alginate are 2–6 × 10^6 cells per milliliter and 0.75–1 % (wt/vol), respectively. After nearly instantaneous gelation, cell spots on the 532 micropillars are immersed in growth medium in the complementary 532 microwells by stamping. The sandwiched DataChip is incubated in a CO_2 incubator at 37 °C for 18 h prior to toxicity tests.

1.3.2 Enzyme and Compound Spotting

The MetaChip is transversely divided into typically four regions (A–D) by printing 120 nL of metabolic enzyme/Matrigel mixtures in the microwell chip laid on a chilling slide deck at 4 °C. Specifically, regions A to D contain no enzyme as a test compound only control, a mixture of human cytochrome P450 isoforms (P450 Mix), a mixture of P450 Mix and human Phase II metabolizing enzymes (All Mix), and human liver microsomes (HLM). Immediately after spotting, the MetaChip is stored in a −80 °C freezer until use. For toxicity tests, six different compounds are printed in regions 1–6 of the MetaChip, creating twenty-four distinct regions on the chip, each region containing a 3 × 6 mini-array. Within each mini-array, six different doses of a compound are assayed for toxicity. Thus, a single Data chip combined with a single MetaChip has the capability to generate twenty-four dose response curves for six compounds and their metabolites generated from three different metabolic conditions on the chip.

1.3.3 Chip Incubation, Staining, Scanning, and Data Analysis

After stamping the DataChip onto the MetaChip and incubating the combined chips for 24 h in the presence of compounds followed by 48 h post-stamping incubation in fresh growth medium, the DataChip containing cells is stained with a live/dead viability/cytotoxicity kit from Life Technologies. Staining dyes such as calcein AM and ethidium homodimer-1 are used to produce a green fluorescent response from live cells and a red fluorescent signal from dead cells. The dried DataChip is scanned with S+ Scanner to obtain fluorescent images of cell spots. Alternatively, a GenePix Professional 4200 A scanner and Cellomics ToxInsight *In Vitro* Toxicity (IVT)

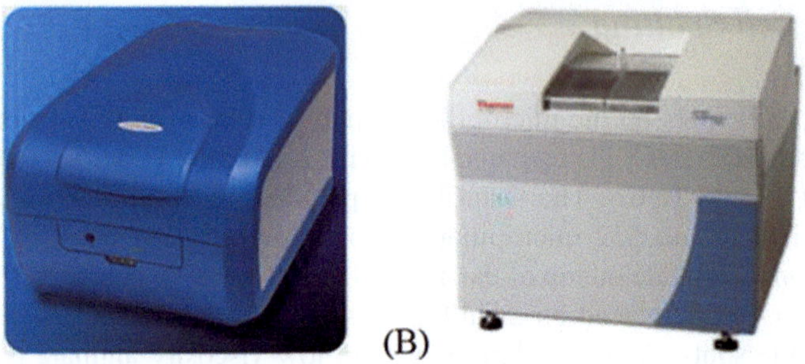

Fig. 1.13 (A) GenePix 4200 A microarray scanner from Molecular Devices and (B) Cellomics ToxInsight *In Vitro* Toxicity (IVT) platform from Thermo Fisher are required to obtain scanned images from the DataChip/MetaChip.

Table 1.5 Comparison of the DataChip/MetaChip platform with the 96-well plate platform

Platform	Spot volume	Morphology	Compound solution volume	Compound solution/cell ratio (nL/cell)	Incubation time (h)
DataChip (532-micropillar chip)	60 nL (200 cells/spot)	3D cells in alginate/Matrigel	–	–	24–72
MetaChip (532-microwell chip)	120 nL (54 nM enzyme)	Metabolizing enzymes in alginate/Matrigel	720 nL	3.6 (720 nL/200 cells)	24
Well plate (96-well plate)	100 μL (10,000 cells/well)	2D cell monolayers	100 μL	10 (100,000 nL/10,000 cells)	24–72

platform can be used (Fig. 1.13). The dose response curves and IC_{50} values are calculated from scanned images using ImageJ and S+ Chip Analysis.

The fundamentally distinctive advantages of the DataChip/MetaChip platform are the result of product and process characteristics designed and engineered to replicate the human metabolism process. Consequently, the platform features a 3D cellular environment that mimics that of human tissues *in vivo*. It has the ability to accommodate a diverse array of cells, metabolizing enzymes and enzyme mixtures, and a broad array of drug candidates. Combined with platform aspects that optimize cell growth, these system characteristics transform to robust competitive strengths of high throughput, high predictive reliability, low cost, and consistent reproducibility of toxicity profiles (Table 1.5).

1.4 Advantages and Applications

The microarray chip technology offers several clear advantages over more conventional *in vitro* toxicology screening tools. Specifically, it requires extremely small amounts of enzymes, viruses, cells, compounds, and reagents for analysis. It is well suited for early stage HTS of compound libraries, which include drug candidates, chemicals, cosmetic ingredients, and environmental toxicants. The cell encapsulation technology developed is flexible and allows one to culture multiple cell types from different organs in hydrogel droplets on the chip, thus providing more predictive insight into potential organ-specific toxicity of compounds. 3D-cell cultures on the chip may provide an environment that simulates the *in vivo* ECM conditions, and therefore help to maintain the specific biochemical functions and morphological features of human cells similar to those found in human organ tissues. Through the use of recombinant adenoviruses, the chip enables the controlled expression of each human metabolizing enzyme as well as various combinations of multiple enzymes in human cells. The gene transduction technology on the chip can be extended to other cell-based *in vitro* assays, including gain- and loss-of-function genomic screening. The chip could be tailored to different subgroups of the population and even to individual patients by adjusting the expression levels of the various metabolizing enzymes in human liver cells to match the enzyme levels representative of a subgroup or unique to an individual. Using this technology, adverse responses of drug candidates and their reactive metabolites by combinations of various DMEs in the human liver can be assessed and quantified at speeds commensurate with early-stage human toxicity tests. Ultimately, this approach would provide critical information needed for the design of patient-specific treatment regimens, as well as for the identification of pharmacologically safe and effective lead compounds for advancement to clinical trials. With "high quality" information on chemical toxicity early on, pharmaceutical companies can make better educated decisions on which compounds to take forward, accelerate drug development times, and reduce investment in late-stage drug failures.

A list of microarray bioprinting-based assays is provided in Fig. 1.14, and more information can be found in Chap. 8.

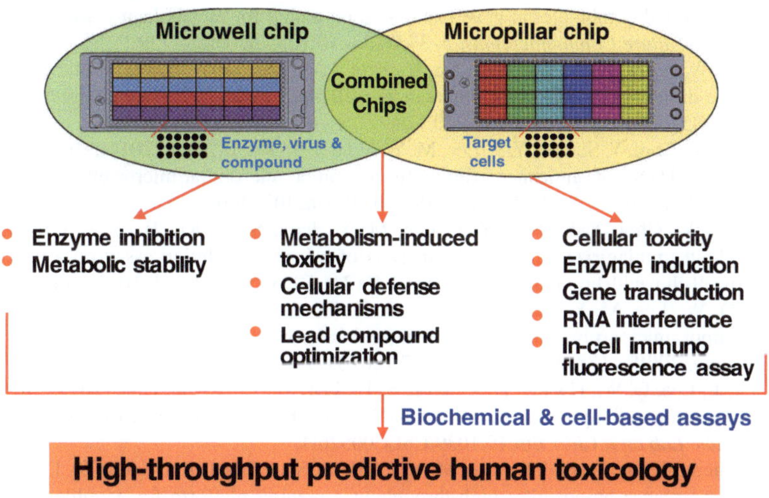

Fig. 1.14 Examples of microarray bioprinting-based assays for high-throughput, predictive toxicology screening and drug discovery.

References

1. Datar, A., Joshi, P., & Lee, M.-Y. (2015). Biocompatible hydrogels for microarray cell printing and encapsulation. *Biosensors, 5*(4), 647–663. doi:10.3390/bios5040647.
2. Joshi, P., & Lee, M. Y. (2015). High content imaging (HCI) on miniaturized three-dimensional (3D) cell cultures. *Biosensors, 5*(4), 768–790. doi:10.3390/bios5040768.
3. Lee, M.-Y., Park, C. B., Dordick, J. S., & Clark, D. S. (2005). Metabolizing enzyme toxicology assay chip (MetaChip) for high-throughput microscale toxicity analyses. *Proceedings of the National Academy of Sciences, 102*(4), 983–987. doi:10.1073/pnas.0406755102.
4. Lee, M. Y., & Dordick, J. S. (2006). High-throughput human metabolism and toxicity analysis. *Current Opinion in Biotechnology, 17*(6), 619–627. doi:10.1016/j.copbio.2006.09.003.
5. Lee, M. Y., Clark, D. S., & Dordick, J. S. (2006). Human P450 microarrays for in vitro toxicity analysis: Toward complete automation of human toxicology screening. *Journal of the Association for Laboratory Automation, 11*(6), 374–380. doi:10.1016/j.jala.2006.08.003.
6. Lee, M.-Y., Dordick, J., & DS, C. (2010). Metabolic enzyme microarray coupled with miniaturized cell-culture array technology for high-throughput toxicity screening. *Methods in Molecular Biology, 632*, 221–237. doi:10.1007/978-1-60761-663-4.
7. Lee, D. W., Lee, M. Y., Ku, B., Yi, S. H., Ryu, J. H., Jeon, R., et al. (2014). Application of the DataChip/MetaChip technology for the evaluation of ajoene toxicity in vitro. *Archives of Toxicology, 88*(2), 283–290. doi:10.1007/s00204-013-1102-9.
8. Ziauddin, J., & Sabatini, D. M. (2001). Microarrays of cells expressing defined cDNAs. *Nature, 411*(6833), 107–110. doi:10.1038/35075114.
9. Lee, M.-Y., Kumar, R. A., Sukumaran, S. M., Hogg, M. G., Clark, D. S., & Dordick, J. S. (2008). Three-dimensional cellular microarray for high-throughput toxicology assays. *Proceedings of the National Academy of Sciences, 105*(1), 59–63. doi:10.1073/pnas.0708756105.
10. Fernandes, T. G., Kwon, S. J., Bale, S. S., Lee, M. Y., Diogo, M. M., Clark, D. S., et al. (2010). Three-dimensional cell culture microarray for high-throughput studies of stem cell fate. *Biotechnology and Bioengineering, 106*(1), 106–118. doi:10.1002/bit.22661.
11. Fernandes, T. G., Kwon, S. J., Lee, M. Y., Clark, D. S., Cabral, J. M. S., & Dordick, J. S. (2008). On-chip, cell-based microarray immunofluorescence assay for high-throughput analysis of target proteins. *Analytical Chemistry, 80*(17), 6633–6639. doi:10.1021/ac800848j.
12. Kwon, S. J., Lee, D. W., Shah, D. A., Ku, B., Jeon, S. Y., Solanki, K., et al. (2014). High-throughput and combinatorial gene expression on a chip for metabolism-induced toxicology screening. *Nature Communications, 5*, 3739. doi:10.1038/ncomms4739.
13. Lee, D. W., Yi, S. H., Jeong, S. H., Ku, B., Kim, J., & Lee, M. Y. (2013). Plastic pillar inserts for three-dimensional (3D) cell cultures in 96-well plates. *Sensors and Actuators B: Chemical, 177*, 78–85. doi:10.1016/j.snb.2012.10.129.
14. Lee, D. W., Choi, Y. S., Seo, Y. J., Lee, M. Y., Jeon, S. Y., Ku, B., et al. (2014). High-throughput screening (HTS) of anticancer drug efficacy on a micropillar/microwell chip platform. *Analytical Chemistry, 86*(1), 535–542. doi:10.1021/ac402546b.
15. Lee, D. W., Choi, Y.-S., Seo, Y. J., Lee, M. Y., Jeon, S. Y., Ku, B., et al. (2014). High-throughput, miniaturized clonogenic analysis of a limiting dilution assay on a micropillar/microwell chip with brain tumor cells. *Small, 10*(24), 5098–5105. doi:10.1002/smll.201401074.
16. Lee, D. W., Lee, M.-Y., Ku, B., & Nam, D.-H. (2015). Automatic 3D cell analysis in high-throughput microarray using micropillar and microwell chips. *Journal of Biomolecular Screening, 20*(9), 1178–1184. doi:10.1177/1087057115597635.
17. Kang, J., Lee, D. W., Hwang, H. J., Yeon, S.-E., Lee, M.-Y., & Kuh, H.-J. (2016). Mini-pillar array for hydrogel-supported 3D culture and high-content histologic analysis of human tumor spheroids. *Lab on a Chip*. doi:10.1039/C6LC00526H.
18. Sukumaran, S. M., Potsaid, B., Lee, M.-Y., Clark, D. S., & Dordick, J. S. (2009). Development of a fluorescence-based, ultra high-throughput screening platform for nanoliter-scale cytochrome p450 microarrays. *Journal of Biomolecular Screening, 14*(6), 668–678. doi:10.1177/1087057109336592.

19. Kwon, S. J., Lee, M. Y., Ku, B., Sherman, D. H., & Dordick, J. S. (2007). High-throughput, microarray-based synthesis of natural product analogues via in vitro metabolic pathway construction. *ACS Chemical Biology, 2*(6), 419–425. doi:10.1021/cb700033s.
20. Zhang, H., Lee, M. Y., Hogg, M. G., Dordick, J. S., & Sharfstein, S. T. (2012). High-throughput transfection of interfering RNA into a 3D cell-culture chip. *Small, 8*(13), 2091–2098. doi:10.1002/smll.201102205.
21. Park, T.-J., Lee, M.-Y., Dordick, J. S., & Linhardt, R. J. (2008). Signal amplification of target protein on heparin glycan microarray. *Analytical Biochemistry, 383*(1), 116–121. doi:10.1016/j.biotechadv.2011s.08.021.Secreted.
22. Kwon, S. J., Mora-Pale, M., Lee, M. Y., & Dordick, J. S. (2012). Expanding nature's small molecule diversity via in vitro biosynthetic pathway engineering. *Current Opinion in Chemical Biology, 16*(1–2), 186–195. doi:10.1016/j.cbpa.2012.02.001.

Chapter 2
Microarray Spotter and Printing Technologies

Akshata Datar, Dong Woo Lee, Sang Youl Jeon, and Moo-Yeal Lee

Contents

2.1 Introduction 20
  2.1.1 Contact Printing Techniques 20
  2.1.2 Non-contact Printing Techniques 21
2.2 Materials 22
2.3 Components of S+ MicroArrayer 22
  2.3.1 Main Body Components 23
  2.3.2 Externally Connected Components 26
2.4 General Precautions for Operating S+ MicroArrayer 27
2.5 Daily Operations of S+ MicroArrayer 28
  2.5.1 Turning 'ON' the System 29
  2.5.2 Refilling the Pressure Bottles 29
  2.5.3 Washing Tubes, Solenoid Valves, and Ceramic Tips with Ethanol and Water 30
  2.5.4 Dispensing Samples on the Micropillar/Microwell Chips with a Work File 32
  2.5.5 Replacing Solenoid Valves and Ceramic Tips 34
  2.5.6 Turning 'OFF' the System 37
2.6 Detailed Programming for Normal Operation 38
  2.6.1 Generating Wash and Dry Sequences 38
  2.6.2 Defining Well Plates 38
  2.6.3 Registering Chips 41
  2.6.4 Registering Spot Layouts 44
  2.6.5 Optimizing Dispensing Parameters Using Vision Inspection 47
2.7 Summary 50
References 50

M.-Y. Lee, *Microarray Bioprinting Technology*,
DOI 10.1007/978-3-319-46805-1_2

2.1 Introduction

The early stages of drug discovery heavily rely on high-throughput screening (HTS) of compound libraries to identify effective lead compounds. Biochemical and cell-based assays have been commonly performed with the assistance of robotic liquid dispensing systems to rapidly test and identify the potential efficacy and toxicity of drug candidates [1, 2]. In an effort to save costs and reduce expensive resources such as primary human cells and reagents, pharmaceutical industries have focused on miniaturizing HTS assays by using higher density well plates including 384- and 1536-well plates, leading to a reduction in reagent volumes and an increase in the speed of the liquid dispensing [1, 3].

Biological sample printing using microarray spotters is an important advancement in the field of miniaturized assay development and HTS. Compared to traditional liquid dispensing systems, microarray spotters allow one to dispense extremely small volumes (typically 200 pL to 950 nL) of biological samples (including reagents, growth media, compounds, hydrogels, genes, proteins, viruses, and cells) in microtiter plates, on glass slides, or on plastic chips. The process of cell printing in hydrogels on a micropillar/microwell chip platform allows a precise positioning of human cell spots, leading to the creation of more physiological relevance of human cells grown in three dimensions (3D) [4, 5].

In this chapter, we will briefly introduce the advantages of various printing technologies and go over general precautions that have to be taken when printing various biological samples with a microarray spotter. In general, microarray spotting is divided into direct contact printing using pins and non-contact printing using solenoid valves, piezoelectric nozzles, *etc.* [6, 7]. Although we have experience in operating several microarray spotters, including MicroSys and PixSys from DigiLab and Nano-Plotter from GeSim, we will provide detailed protocols on how to operate S+ MicroArrayer from Samsung Electro-Mechanics, Co. (SEMCO). S+ MicroArrayer represents the most advanced microarray spotter for cell printing, which is specifically designed to accommodate the micropillar/microwell chip. The principle of operating and troubleshooting S+ MicroArrayer can be applicable to any solenoid-driven microarray spotters.

2.1.1 Contact Printing Techniques

Contact printing technologies mainly function on the principle of transferring biomaterials via directly contacting tips to the surface of functionalized glass slides. Examples of contact printing technology are microarray spotters with an array of multiple contact pins and atomic force microscopes (AFMs) with scanning tips (also known as dip-pen nanolithography). The contact microarray spotters have a single pin or a pin array to transfer liquids from source plates, which is usually a 96- or 384-well plate, to destination glass slides. A droplet forms at the tip when the tip comes in direct contact with a liquid. Several types of tips have been developed, depending on the material used and the capillary size inside the tip to manage the volume of the droplet printed [8]. Factors that determine the volume of the droplet are the capillary size, surface tension of the liquid, affinity of the liquid with the glass slide, and the surface chemistry of the glass slide. Pin-based printing is mainly used to print extremely small volumes of proteins and DNA on the surface of glass slides [9, 10]. Although pin-based printing is straightforward and fast, it is difficult to print colloidal suspensions (such as cells), and the droplet size may be inaccurate and inconsistent depending on surface and liquid properties [9]. For dip-pen nanolithography, AFM is used to deposit some biological samples, mostly small molecules on the surface [7, 11]. Although this technology has capable of printing small spots in 25–200 nm resolution, it is not widely used in printing biological samples due to lack of multiplexing capabilities and limited detection methods [7, 9].

2.1.2 Non-contact Printing Techniques

In non-contact printing, microarray spotters with solenoid valves are the most commonly used, which function on the principle of electromagnetic induction. Typical solenoid valves consist of a metal rod that is surrounded by coiled wires that conduct an electric current. When voltage is applied across the coils, a magnetic field is generated that forces the metal rod to act as a shaft that moves up and down. This metal rod acts as a gate while printing biological samples. The setup is enclosed in a nozzle that has constant sample supply and constant pressure maintained using syringe pumps. When the shaft moves up and down, the outlet for printing biological samples opens and closes. This is how the voltage applied to the solenoid valves controls dispensing of liquid samples [6]. Unlike other dispensing systems, the solenoid valves can handle colloidal samples quite well. Typical dispensing volumes are 30–950 nL per droplet. One particular printer, S+ MicroArrayer, is equipped with six solenoid valves for printing six samples simultaneously.

Piezoelectric nozzles that are used in conventional inkjet printers are also commonly used in microarray printing with dispensing volume of 200–600 pL per droplet. Unlike solenoid valves, the dispensing volume is controlled by the inner structure of piezoelectric nozzles. Microarray spotters with piezoelectric nozzles use electric pulses that control the expansion and contraction of a piezoelectric membrane that acts as a pump to push biological samples through the tip [12, 13]. Thus, the dispensing volume can be increased not by changing voltages, but by depositing multiple spots at the same location. Although piezoelectric nozzles are capable of printing biological samples, its use is limited due to inconsistent printing of colloidal samples and difficulty of controlling large dispensing volumes [14]. Piezoelectric nozzles are more commonly used for printing compounds dissolved in DMSO, proteins, and amino/nucleic acids.

Another popular type of bioprinting technology adapted in the field of tissue engineering is laser-assisted bioprinting (LAB), also known as laser-induced forward transfer (LIFT) [15]. Typically, LIFT consists of a pulsed laser beam, a glass slide coated with a biological sample, and a receiving glass slide. The laser beam is focused onto the absorbing layer, from which the heat is transferred to the biological sample where a bubble is generated. The bubble expands until its explodes, and the material in the bubble gets deposited onto the receiving glass slide [16]. However, this type of printing needs delicate control of the laser spotting, intensity, moisture and viscosity of the sample, thickness of the absorbing layer, and the printing speed so that cell viability cannot be compromised [17].

Acoustic printers became popular when single cell-based experiments needed to be carried out and demand for pL-sized droplets increased [18]. Acoustic printing requires a source for ultrasound generation, a pool of a biological sample, and a receiver glass slide. High intensity ultrasound is generated which is focused on air-liquid interface. When the energy exceeds the surface tension of the sample, liquid is ejected from the pool of the sample [19]. The advantage of this technique is its independence of nozzles and tubing. This technology gains in popularity for printing compounds in DMSO. However, high intensity ultrasound in some cases may cause an obstruction [20]. For example, printing delicate primary cells could be problematic because high vibration energy may damage cell membranes and lower cell viability.

2.2 Materials

- Solenoid valve (Lee Company)
- S+ MicroArrayer with six solenoid valves and ceramic tips (Samsung Electro-Mechanics, Co. or SEMCO, Suwon, South Korea)
- Micropillar and microwell chips (SEMCO, Suwon, South Korea)
- Ceramic tips with 150 μm orifice (SEMCO, Suwon, South Korea)
- Reagent alcohol 200 proof ACS grade (VWR)

2.3 Components of S+ MicroArrayer

To better explain the operation of the microarray spotter, we will classify it under mechanical and software components. Although both components work hand-in-hand, their maintenance and parameter setting are different from each other. Mechanical components are divided into the main body, the utility body, and externally connected parts that play an important role in chilling, rinsing, and sample dispensing (Figs. 2.1 and 2.2).

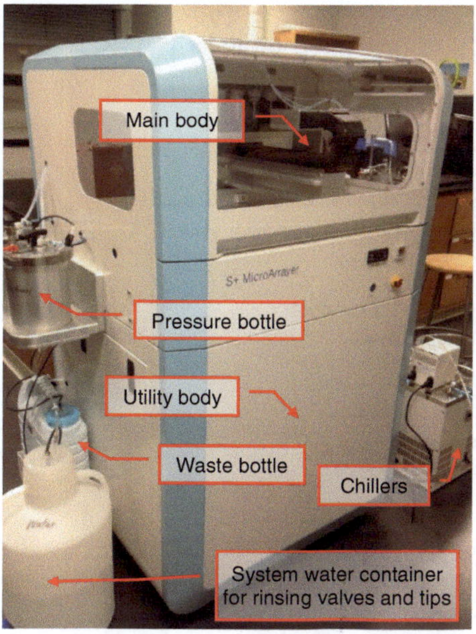

Fig. 2.1 The picture of S+ MicroArrayer

2.3.1 Main Body Components

The main body components are housed in a chamber where biological samples are loaded in a 96-well plate and printed on a micropillar/microwell chip at controlled temperature and humidity. It mainly consists of a robotic arm, a dispensing head with six solenoid values and ceramic tips, six syringe pumps, a droplet inspection camera, a chip inspection camera, two chip-loading decks, a well plate deck, a vacuum pump, a humidifier, a water bath with a sonicator, and a waste drainage basin (Fig. 2.3).

- **Dispensing head unit**

  The dispensing head unit of the microarray spotter is mainly responsible for moving and printing biological samples in accurate volumes and positions. It consists of a chip alignment inspection camera, six solenoid valves connected to syringe pumps through tubing, and six ceramic tips (Fig. 2.4). It is capable of moving X, Y and Z directions, hence aspirating biological samples from 96-well/384-well plates and dispensing in different patterns on the micropillar/microwell chip.

  - **Solenoid valves**: They are the main component behind microarray bioprinting technology, which function on the principle of electromagnetic induction. The voltage applied to the valves, creates a magnetic field that forces the gate to open and close. The syringe pumps maintaining the pressure help the samples to flow when the gate is open. Each solenoid valve can be controlled individually.

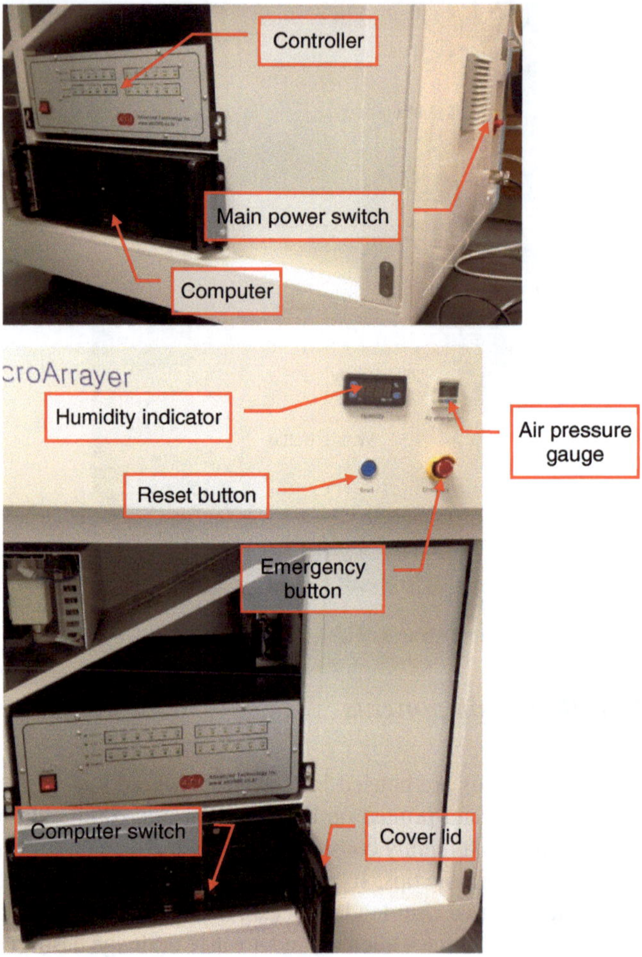

Fig. 2.2 The utility body components of S+ MicroArrayer

  - **Ceramic tips**: They are necessary to dispense small droplets of biological samples when the gate is open. They are made of an inert material for rinsing with alcohol and sonication, hence preventing contamination while several biological samples are printed.
  - **Chip alignment inspection camera**: It checks the position of micropillar/microwell chip on the chilled chip-loading deck for accurate sample dispensing.

- **Syringe pumps**: The main role of the six syringe pumps is to maintain pressure when the samples need to be aspirated or dispensed. When the gates of the six solenoid valves open to dispense the samples, the positive pressure main-

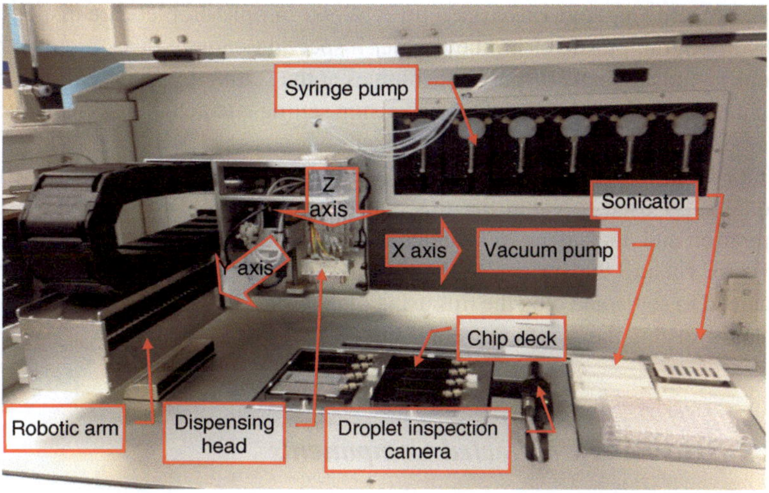

Fig. 2.3 The main body components of S+ MicroArrayer

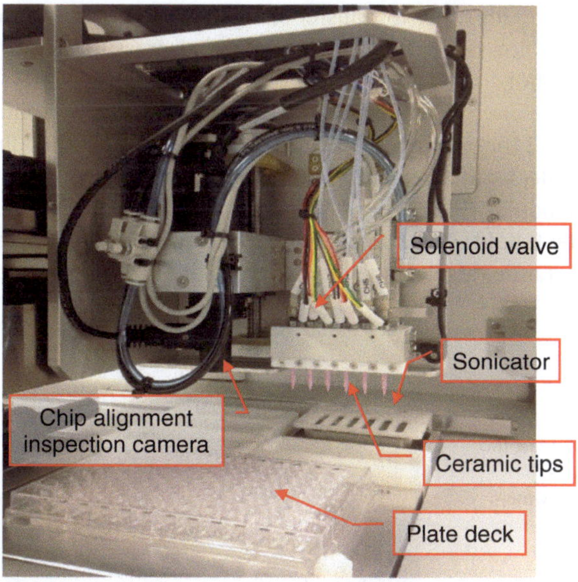

Fig. 2.4 The dispensing head unit of S+ MicroArrayer

tained by the syringe pumps forces the samples out on the micropillar and microwell chips.

- **Droplet inspection camera**: There are two cameras in the main body of the microarray spotter. The chip alignment inspection camera is installed in the dispensing head, and the droplet inspection camera is placed beside the chip-loading deck, which is used to optimize dispensing parameters such as air pressure and solenoid valve open time. When sample droplets are dispensed on a hydrophobic plastic strip, the droplet inspection camera takes pictures of the droplets and calculate dispensing volumes at the setting by measuring the height and diameter of the droplet. The dispensing setting and actual dispensing volume can be compared for parameter optimization.

- **Water bath with a sonicator and a vacuum pump**: These two components are sequentially used to clean the tips and tubing that hold the samples between aspiration and dispensing of samples. The water bath with a sonicator is responsible for rinsing the surface of the tips and breaking down any bigger particles that tend to settle and clog the tips using ultrasound. The vacuum pump is responsible for drying the tips to prevent sample carry-over.

2.3.2 Externally Connected Components

- **Pressure bottles**: One pressure bottle contains water for sample printing and the other bottle holds 70 % ethanol for rinsing and sterilization (Fig. 2.5). The tubing has to be filled with water to transfer energy from the syringe pumps and push the samples to flow for printing. The bottles are pressurized with air to supply water and alcohol.
- **Chillers**: There are two chillers equipped in the microarray spotter, one for the dispensing head for printing temperature-sensitive samples and the other for the chip-loading decks for preventing spot drying, the 96-well plate deck, and the two pressure bottles (Fig. 2.6).
- **Waste bottle**: It holds wastewater generated by rinsing and washing the solenoid valves, the ceramic tips, and the tubing.

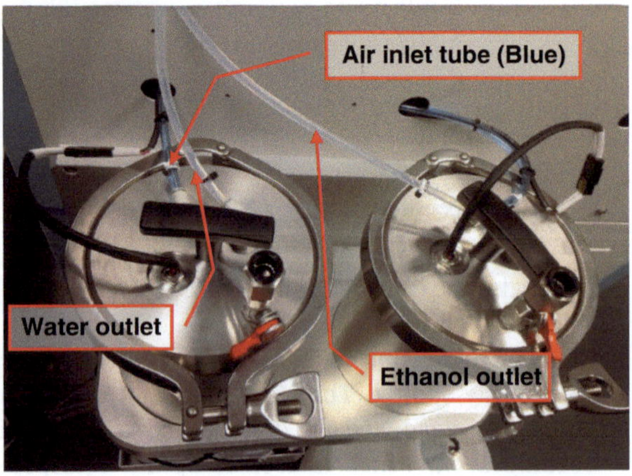

Fig. 2.5 The picture of the two pressure bottles containing distilled water and 70 % ethanol

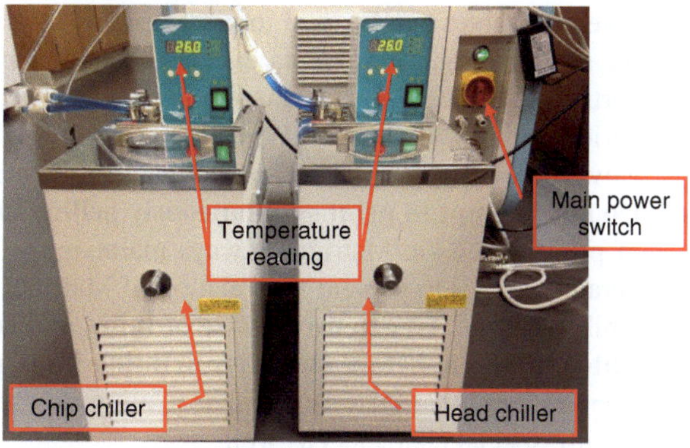

Fig. 2.6 The picture of the two chillers for preventing water evaporation on the chip and printing temperature-sensitive samples

2.4 General Precautions for Operating S+ MicroArrayer

- The S+ MicroArrayer uses surface cooling and condensation so that it requires low humidity setting. In particular, water condensation on the moving part (not on the chip-loading deck) or the circuit board would be detrimental. **Do not increase relative humidity higher than 60 % at room temperature.** In addition, maintain the temperature of the chip-loading deck between **4–10 °C** to retard evaporation of water in spots on the chip. **It is extremely important to avoid excess water condensation on the bottom layer (e.g., $BaCl_2$/PLL spots) to prevent spot detachment.**
- Unlike MicroSys and PixSys microarray spotters, there is no vacuum applied to hold the chips on the S+ MicroArrayer. **Before spotting, make sure that all chips lie flat on the spring-loaded chip deck and no obstacles are on the work place.** It is the main cause of the ceramic tips crashed.
- **All ceramic tips should be positioned inside of 96-wells when aspirating samples**. If any one of the tips left outside of the 96-well plates when aspirating due to wrong programming, then they will be broken. **Always use the same kind of the 96-well plates to avoid tip crashing due to Z height difference in 96-wells.**
- Washing solenoid valves and ceramic tips with alcohol is essential to remove air bubbles prior to sample dispensing. **After alcohol washing, thorough rinsing solenoid valves and ceramic tips with water is necessary to avoid enzyme deactivation or cell death due to remaining alcohol.**
- **Do not change Z levels randomly to prevent the ceramic tips crashed into the chip surface.** The typical distance between the chip surface and the tip is approximately **0.7 mm**.
- Always check a streamline of water while rinsing solenoid valves. If the streamline is deflected from the vertical or water is beaded up at the ceramic tip end, clean the ceramic tip with sonication. It is highly important to stop working with a clogged tip immediately.

- Solenoid valves can be used to print viscous solutions, but may not be suitable for spotting organic solvents except alcohols because of incompatible plastic parts in the solenoid valve. Aspirating DMSO with solenoid valves for a short period of time will be okay. When compounds in DMSO is diluted with solenoid valves, use a 384-well plate, and do not aspirate more than 20 μL of samples.
- **Do not attempt to print spontaneously gelling materials in any case.** In case of printing cells in Matrigel, always maintain the dispensing head at low temperature (4 °C) to prevent the gelation of Matrigel inside of the solenoid valves while printing. **An extensive rinse of tubes and solenoid valves with ice-cold water is critical before and after printing Matrigel solution.**
- Keep in mind that there is no "space" allowed when saving a file name, and do not use special characters as well.
- If you have a new operational file made, test it without ceramic tips installed first. Do not install the expensive tips for testing unknown work files.

2.5 Daily Operations of S+ MicroArrayer

Now that we understand the mechanical components of the microarray spotter, it is important to know how to control these components and make them work as a whole. ezAOI is a customized software developed to operate S+ MicroArrayer (Fig. 2.7). This user interface helps us to control the microarray spotter and print

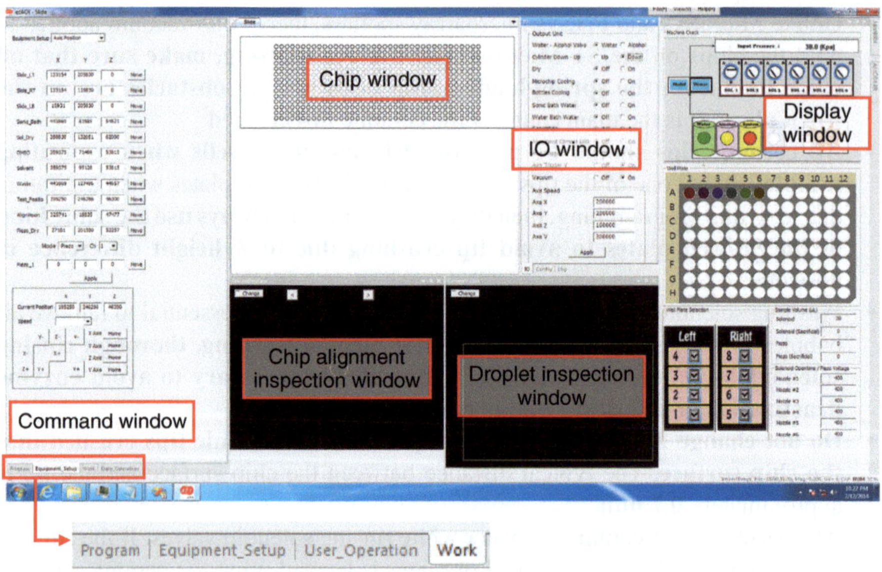

Fig. 2.7 The main screen of ezAOI. The Command window has four groups—Program, Equipment setup, User operation, and Work

samples on the micropillar and microwell chips in distinctive patterns. The main screen that pops up displays six windows, including command window, slide window, IO window, display window, chip alignment inspection window, and droplet inspection window. The command window in itself has four tabs, including program, equipment setup, user operation, and work. Each of these tabs has different functions, which allow us to program the microarray spotter more efficiently. For daily operation of the microarray spotter, follow the protocols provided below.

2.5.1 Turning 'ON' the System

1. Open the air cylinder (or in-house air valve) and maintain the pressure of compressed air at 100 psi.
2. Turn on the chiller. ***Note: For most of applications, there is no need to turn on the humidifier because of surface cooling. Maintain the temperature of the chip-loading deck between 4–10 °C to reduce evaporation of water in spots on the chip. It is extremely important to avoid excess water condensation on the bottom layer to prevent spot detachment.***
3. Turn on the external switch in the utility body.
4. Turn on the computer and the monitor.
5. Prior to running the 'ezAOI', push 'Reset button' and reset XYZ coordinates. ***Note: This step is essential to avoid malfunctioning of the microarray spotter. Do not skip this step!***
6. Run the 'ezAOI' software. The software will initialize the system automatically when running. ***Note: Make sure that no obstacles are on the work place to avoid the robotic arm crashing.***

2.5.2 Refilling the Pressure Bottles

1. Select the 'User Operation' window.
2. Go to 'Water Alcohol Change' in the 'Daily Operation' box (Fig. 2.8).
3. Click 'Release' in the 'Air pressure' box to release air pressure in the pressure bottles.
4. Turn the orange knob connected to the pressure bottles from 'S' to 'O' position, which releases the pressure inside.
5. Open the lid of the pressure bottles and fill the pressure bottles with distilled water and 70 % ethanol. ***Note: Make sure to have no precipitates at the bottom of the pressure bottles. Dust and precipitates can be the source of solenoid valve and tip clogging. Occasionally, the pressure bottles have to be cleaned and sterilized. To minimize microbial contamination, distilled water with Clear Bath® or sterilized distilled water can be used.***

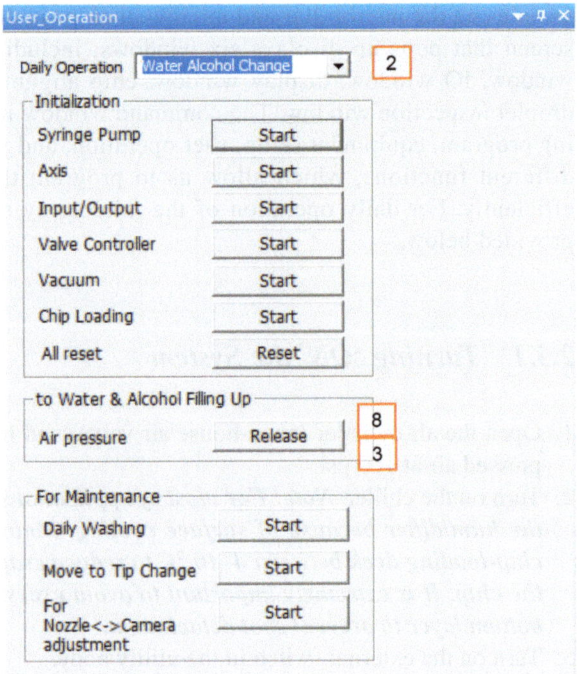

Fig. 2.8 The screen of 'User Operation' > 'Water Alcohol Change'

6. Close the lid of the pressure bottles carefully. ***Note: Make sure that the 'O' ring is properly placed before placing the lid.***
7. Turn the orange knob of the pressure bottles from 'O' to 'S' position, which close the pressure bottles completely.
8. Click 'Initialize' of the 'Air pressure' box in the 'Water & Alcohol Filling Up' box to apply air pressure in the bottles.

2.5.3 Washing Tubes, Solenoid Valves, and Ceramic Tips with Ethanol and Water

1. Select the 'User Operation' window.
2. Select 'Wash & Dry' in the 'Daily Operation' box (Fig. 2.9).
3. Select the wash and dry sequence you generated.
4. Select the number of solenoid valves you want to wash and dry in the 'Select Nozzle' box.
5. By clicking the 'Run Wash & Dry' button, run the wash and dry sequence. ***Note: Washing tubes and solenoid valves with alcohol is essential to remove air bubbles prior to sample dispensing. After alcohol washing, thorough rinsing tubes and solenoid valves with water is necessary to avoid enzyme deactivation or cell death due to remaining alcohol. An extensive rinse of tubes***

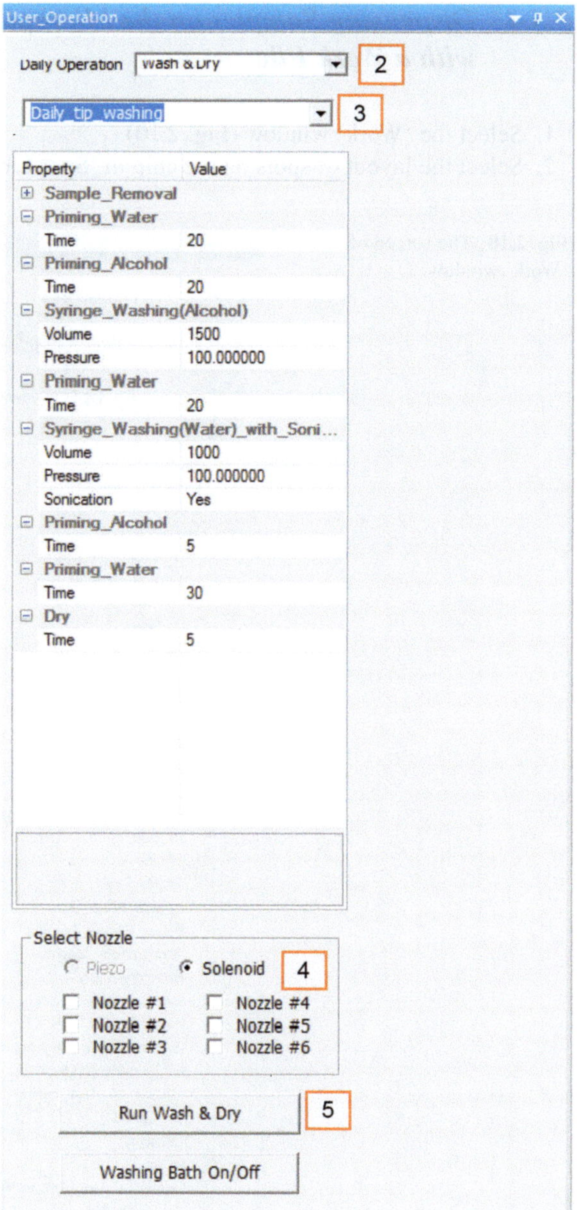

Fig. 2.9 The screen of 'User Operation' > 'Wash & Dry'

and solenoid valves with ice-cold water is critical before and after printing Matrigel solutions.

6. Check a streamline of water while washing. ***Note: If the streamline is deflected from the vertical or water is beaded up at the tip end, clean the ceramic tip with sonication.***

2.5.4 *Dispensing Samples on the Micropillar/Microwell Chips with a Work File*

1. Select the 'Work' window (Fig. 2.10).
2. Select the layout of spots on the chip in 'Spot Layout'.

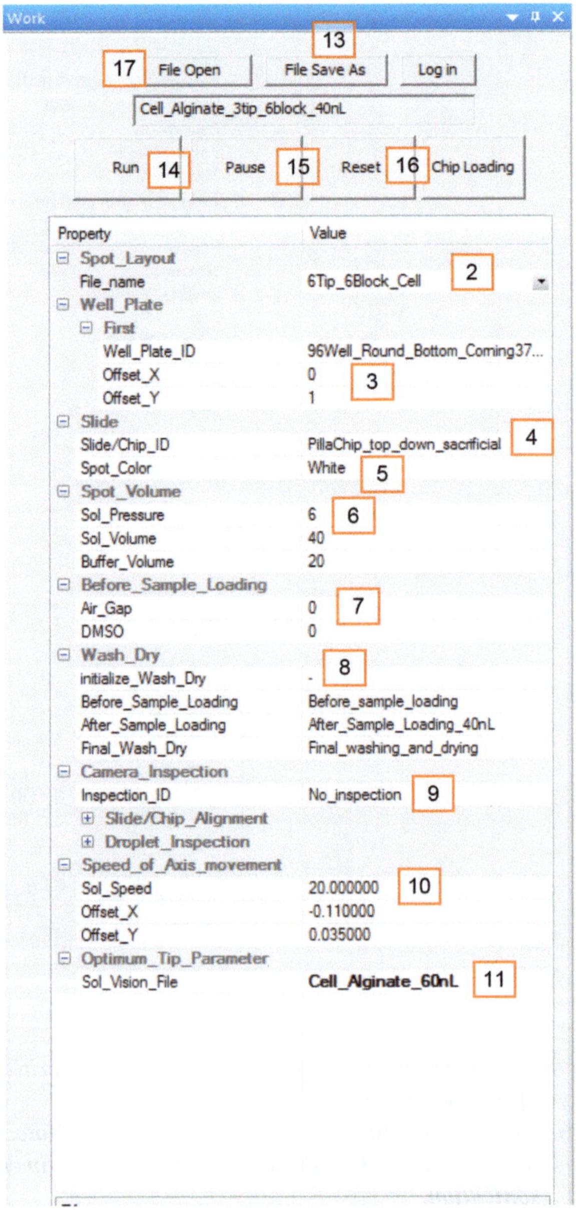

Fig. 2.10 The screen of 'Work' window

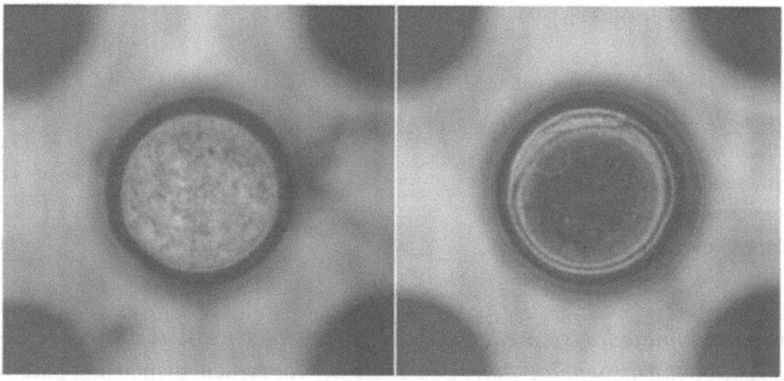

Fig. 2.11 The images of the original micropillar shown as "white spot" (*left*) and the micropillar with the $BaCl_2$-PLL bottom layer shown as "black spot" (*right*)

3. Select the well plate used for sample loading and aspiration in 'Well Plate ID'. ***Note: It is extremely important to use the same kind of the 96-well plates always to avoid tip crashing due to Z height difference in 96-wells.*** Enter the offset values in the X and Y directions to designate the location of samples in 96-wells. The exact location of sample wells will be appeared in the schematic of the well plate in the 'Display' window.
4. Select the type of a chip in 'Slide/Chip ID' onto which samples are dispensed. The schematic of the chip selected will be appeared in the 'Chip' window.
5. Select the color of spots on the chip under the camera in 'Spot Color'. Typically, it is white when nothing is dispensed on the chip or black when $BaCl_2$/PLL is dispensed (Fig. 2.11). ***Note: This color selection is necessary to automatically identify the location of the chip with the chip alignment inspection camera.***
6. Enter the air pressure used for sample spotting and the desired droplet volume for solenoid valves in 'Spot Volume'. ***Note: The typical air pressure used is 6 kilopascal (kPa) for 40–950 nL droplets.***
7. Enter the air gap between the sample and water. ***Note: The typical air gap used is 0 μL for 40–150 nL droplets and 20 μL for 700–950 nL droplets. For accurate sample spotting, small or no air gap is allowed for small dispensing volumes.***
8. Select the wash and dry sequences you want before and after sample dispensing in 'Wash Dry'.
9. Select 'No inspection' in 'Camera Inspection ID'. ***Note: For rapid sample dispensing, we typically skip camera inspection of droplets. Camera inspection of sample spotting is necessary to optimize dispensing parameters at first.***
10. Enter the speed of the axis movement for solenoid valves in 'Speed of Axis Movement'. It is typically 10–20 mm/s.
11. Open the optimum dispensing parameters of solenoid valves obtained from the vision inspection in 'Optimum Tip Parameter'. ***Note: Manually change the open time, if needed. Optimum dispensing parameters will not be changed for each sample and droplet volume used, unless the solenoid valves are broken***

Table 2.1 Typical dispensing parameters for solenoid valves

Samples printed	Spot volume (nL)	Air pressure (kPa)	Air gap (μL)	Open time (μs)
$BaCl_2$/PLL	60	6	0	600
Cells in alginate	60	6	0	600
Enzymes in Matrigel	100	6	0	600
Virus in growth media	320	6	5	4500
Growth media	950	6	20	12,000

or contaminated. Use predetermined parameters without droplet inspection to reduce the dispensing time. See Table 2.1 for typical dispensing parameters, including air pressures, air gaps, and solenoid valve open times.

12. Place the chips on the chip-loading deck and select the location of chips in the 'Display' window onto which samples are dispensed. ***Note: It is extremely important not to leave the chip unattended long time on the chilling chip deck because excess water condensation on the bottom layer (e.g., $BaCl_2$/PLL spot) on the chip will facilitate cell spot detachment.***
13. Click the 'File Save' button and enter a file name to save all information. ***Note: There is no space and special character allowed in the file name.***
14. Add proper amounts of samples in the designated 96-wells shown in the 'Display' window, place the 96-well plate on the deck, and then dispense samples on the chips by clicking the 'Run' button.
15. Click the 'Pause' button to pause the work process. ***Note: In case of emergency such as robotic arm and tip crashing, push the red 'Emergency' button to stop the process (Fig. 2.2).***
16. Click the 'Reset' button to reset the work process after clicking the 'Pause' button.
17. When dispensing samples with a saved work file, click the 'File Open' button, select the work file, place the chips on the chip-loading deck, select the location of desired chips to print in the 'Display' window, add proper amounts of samples in the designated 96-wells, place the 96-well plate on the plate deck, and then click the 'Run' button to print the samples on the chips.

2.5.5 Replacing Solenoid Valves and Ceramic Tips

In case of ceramic tips clogging and solenoid valves malfunctioning, ceramic tips and solenoid valves have to be removed and replaced. Prior to removing tips and solenoid valves, run 'Solenoid Fix', force to dispense large droplets multiple times (typically 500 times), and rinse solenoid valves with water and sonication.

1. Select the 'User Operation' window.
2. Select 'Wash & Dry' under Daily Operation options (Fig. 2.12).
3. Select 'Solenoid Fix' and select all the nozzles from 1 through 6.

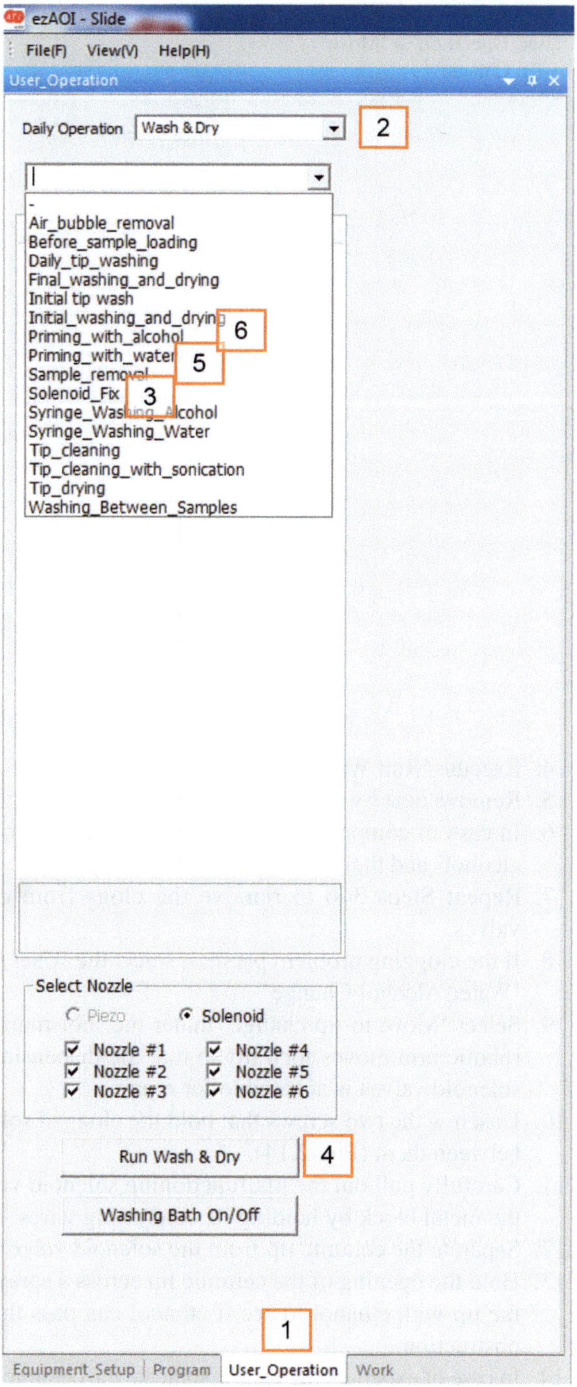

Fig. 2.12 The screen of 'User Operation' > 'Wash & Dry'

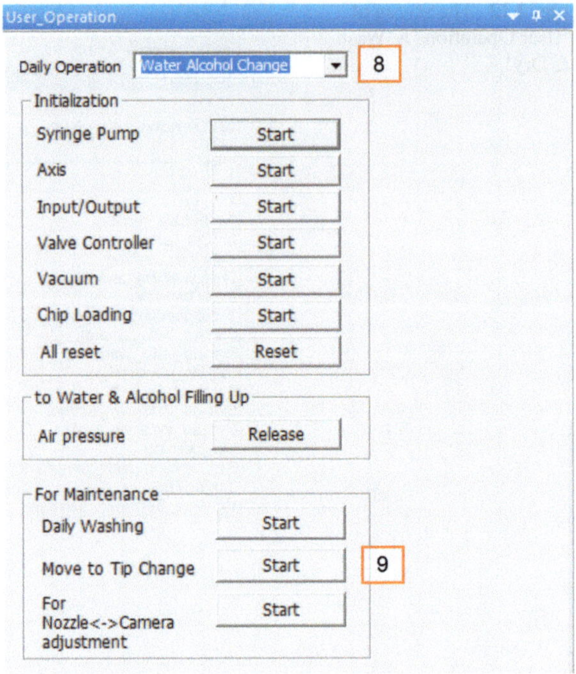

Fig. 2.13 The screen of 'User_Operation' > 'Move to Tip Change'

4. Execute 'Run Wash & Dry'.
5. Remove dust by running 'Priming with water'.
6. In case of compound precipitates clogging solenoid valves, run 'Priming with alcohol' and then execute 'Run Wash and Dry'.
7. Repeat Steps 3–6 to remove the clogs from ceramic tips and/or solenoid valves.
8. If the clogging problem persists, select the 'User Operation' window and select 'Water Alcohol Change'
9. Select 'Move to tip change' under the 'For maintenance' tab (Fig. 2.13). The robotic arm moves forward so that the dispensing head with ceramic tips and solenoid valves is accessible for repair.
10. Unscrew the two screws that hold the clogged solenoid valve or the clogged tip between them (Fig. 2.14).
11. Carefully pull out the malfunctioning solenoid valve with the ceramic tip from the metal block by holding its connecting wires.
12. Separate the ceramic tip from the solenoid valve by gently pulling the tip.
13. Hold the opening of the ceramic tip across a spray bottle with ethanol and flush the tip with ethanol to see if ethanol can pass through the orifice without any obstruction.
14. In case of experiencing a clog, sonicate the ceramic tip until dust is removed. If the clogging problem persists, replace the ceramic tip.

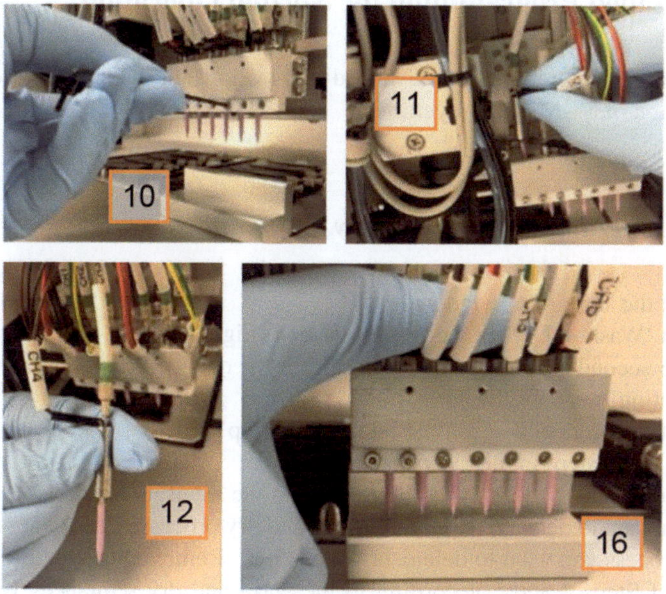

Fig. 2.14 The pictures of the dispensing head for replacing ceramic tips and solenoid valves

15. Attach a new ceramic tip to the solenoid valve, place them accordingly, and connect the wire back in.
16. Before putting the screws in, make sure that the ceramic tips are leveled on the metal block.
17. If the new ceramic tip still doesn't print samples properly, the solenoid valve is clogged, thus replacing it with a new one. ***Note: Run 'Daily Washing' after replacing solenoid valves and before printing samples.***

2.5.6 Turning 'OFF' the System

1. To avoid potential contamination issues, rinse tubes and solenoid valves with ethanol and water thoroughly by running 'Daily Washing'.
2. Close 'ezAOI' software.
3. Turn off the computer and the monitor.
4. Turn off the external power switch in the utility body.
5. Turn off the chillers. ***Note: Make sure to turn off the chillers to avoid excess water condensation on the chip-loading deck and the dispensing head, causing a short circuit by water.***
6. Close the air cylinder (or in-house air valve). ***Note: Do not close the air cylinder while 'ezAOI' is on.***

2.6 Detailed Programming for Normal Operation

These sections are prepared for advanced users who want to know how to program the operation of S+ MicroArrayer.

2.6.1 Generating Wash and Dry Sequences

1. Select the 'Program' window.
2. Select 'Wash & Dry' in the 'Program' box (Fig. 2.15).
3. In the second tab select the wash and dry sequence you made in the 'Daily Operation' window.
4. Select the process you want to add in the 'Step' box.
5. Click 'Add Step' in the 'Step' box.
6. If you want to start all over, click 'New' in the 'Step' box.
7. If you want to delete one of many wash and dry processes you generated, select the process you want to delete, and then click 'Remove Selected Step'.
8. Repeat Steps 4 and 5 according to the wash and dry sequences you want to generate.
9. Change numbers and options in the 'Value' column as you need (Fig. 2.15 & Table 2.2).
10. Click the 'Save as' button to generate the wash and dry sequence.
11. Enter a file name. ***Note: There is no space and special character allowed in the file name.***
12. If you want to modify the wash and dry sequence you generated, select the file name in the box above the 'Save as' button and repeat Steps 8 and 9 as you need.
13. Click the 'Save' button to overwrite the changes in the same file.

2.6.2 Defining Well Plates

1. Select the 'Program' window.
2. Select 'Well Plate' in the 'Program' box.
3. Select the type of a well plate, 96 or 384 (Fig. 2.16).
4. Enter the well layout such as the number of wells and the well-to-well distance (in mm) in X and Y directions. Refer to the 96-well plate figure for correct X and Y directions (Fig. 2.17).
5. Select the 'Equipment Setup' window.
6. Select 'Axis Position' in the 'Equipment Setup' box (Fig. 2.18).
7. Move X, Y and Z axes to locate the first solenoid valve/ceramic tip at the bottom of the A1 well (i.e., the tip locating approximately 3 mm above from the bottom of the well) and then read the current position of X, Y and Z axes.

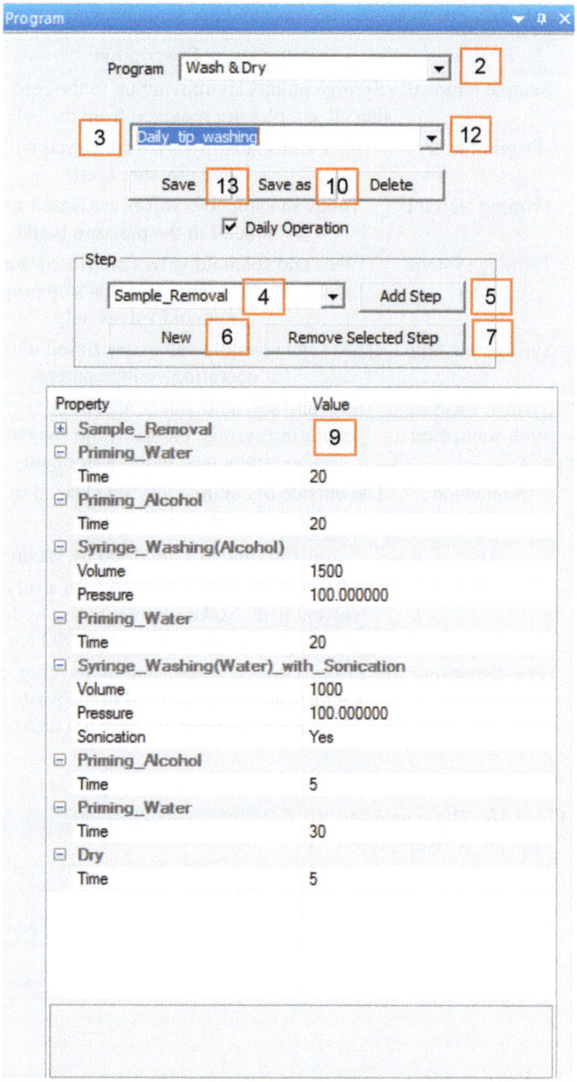

Fig. 2.15 The screen of 'Program' > 'Wash & Dry'

Note: This will be the position of the first ceramic tip for aspirating samples. Do not change Z levels randomly to prevent the tips crashed into the well surface.

8. Select 'Well Plate' in the 'Program' box again.
9. Enter 'Well A1 Position' determined from Step 7.
10. Enter minimum and maximum sample volumes allowed in either 96 or 384 wells.
11. Click 'Save as' button to register well plate information.
12. Enter a file name. ***Note: There is no space and special character allowed in the file name.***

Table 2.2 Processes for the wash and dry sequences

Process	Description	Property	Unit
Sample removal	Syringe pumps are moving up to the zero position so that all samples are removed from the solenoid valves	–	–
Priming water	Tubes and solenoid valves are rinsed with water in the pressure bottle	Time	s
Priming alcohol	Tubes and solenoid valves are rinsed with 70 % ethanol in the pressure bottle	Time	s
Priming system water	Tubes and solenoid valves are rinsed with system water in the big container. This step is applied for solenoid valves only	Time	s
Syringe washing	Tubes and solenoid valves are rinsed with water by operating syringe pumps	Volume	μL
Syringe washing with sonication	Tubes and solenoid valves are rinsed with water by operating syringe pumps while sonicating tips immersed in the water bath	Volume Sonication	μL Yes/No
Sonication	The outside of ceramic tips are cleaned by sonication in the water bath	Time Sonication	s Yes/No
Dry	Ceramic tips are dried by vacuum	Time	s
Solvent wash	Tubes and valves are washed with a solvent in the solvent bath. Make sure not to use hydrophobic solvents. Only alcohols and DMSO are allowed	Volume	μL
Pre-dispensing	Prior to sample dispensing on the chip, aspirated samples are pre-dispensed in the waste drainage basin to equilibrate pressure difference	Droplet	Number

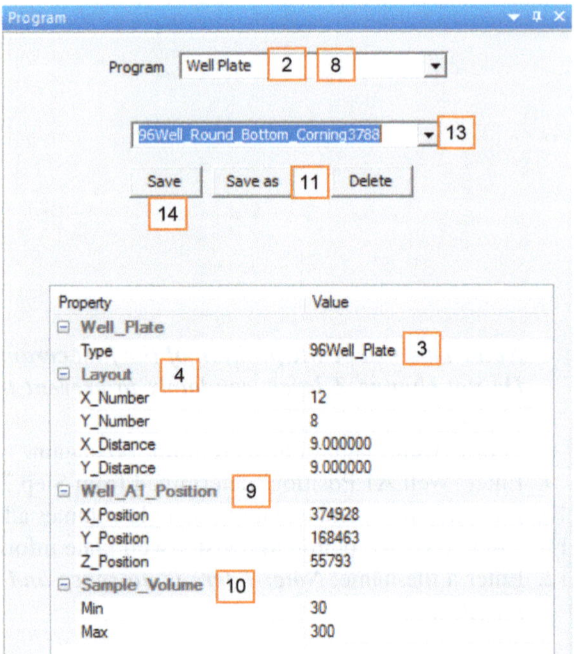

Fig. 2.16 The screen of 'Program' > 'Well Plate'

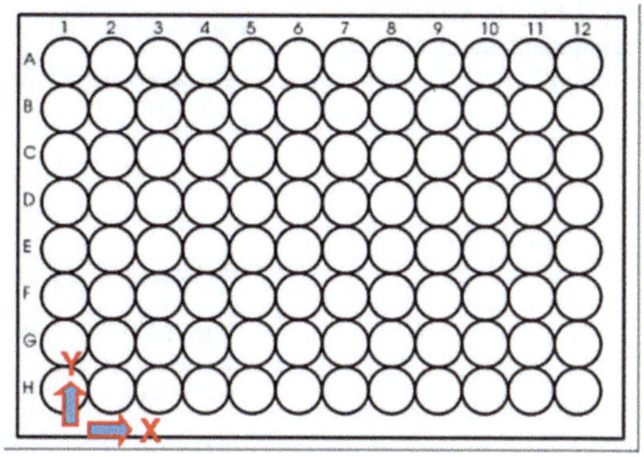

Fig. 2.17 The X and Y directions in the 96-well plate

13. If you want to modify the well plate you registered, select the file name in the box above the 'Save as' button and repeat Steps 3, 4, 9, and 10.
14. Click the 'Save' button to overwrite the changes in the same file.

2.6.3 Registering Chips

1. Select the 'Program' window.
2. Select 'Chip' in the 'Program' box (Fig. 2.19).
3. Enter the information of the chip such as a width (in mm), a height (mm), a margin (mm) from the top left corner, the number of blocks on the chip, a block-to-block distance (mm), the number of spots in each block, a spot-to-spot distance (mm) in X and Y directions. Refer to the slide specification figure for correct X and Y directions (Fig. 2.20).
4. Enter the number of sacrificial spots. For example, 'X distance' 1 means that there is each one of the sacrificial column inserted on the left and right side of the chip, whereas 'Y distance' 1 indicates that there is each one of the sacrificial raw inserted on the top and bottom side of the chip.
5. Check the 'Summary information' of the chip such as the chip margin and the number of spots.
6. Click 'Summary' button to see the schematics of the chip you generated.
7. Select the 'Equipment Setup' window.
8. Select 'Axis Position' in the 'Equipment Setup' box (Fig. 2.21).
9. Move X, Y and Z axes to locate the ceramic tip approximately 0.7 mm above the chip surface and then read the current position of the Z axis.
10. Select 'Chip Inspection Camera' in the 'Equipment Setup' box.
11. Select the box next to 'Camera Live' in the 'Camera Inspection' box (Fig. 2.21).
12. Move X and Y axes to locate the align inspection camera above the chip surface. Move the Z axis to bring the camera into focus and then read the current position of the Z axis.

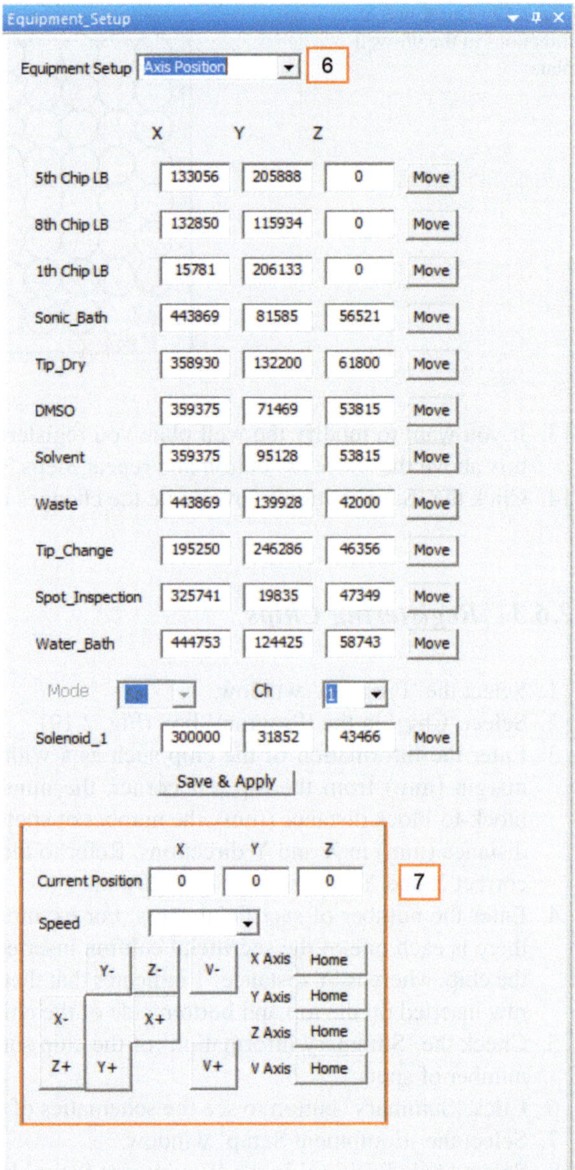

Fig. 2.18 The screen of 'Equipment Setup' > 'Axis Position'

13. Deselect 'Camera Live' in the 'Camera Inspect' box.
14. Select 'Chip' in the 'Program' box again.
15. Enter the Z position of 'Alignment Camera' determined from Step 12 (Fig. 2.19).
16. Enter the Z position of 'Dispensing Height' determined from Step 9.

Fig. 2.19 The screen of 'Program' > 'Chip': (**A**) top and (**B**) bottom panel.

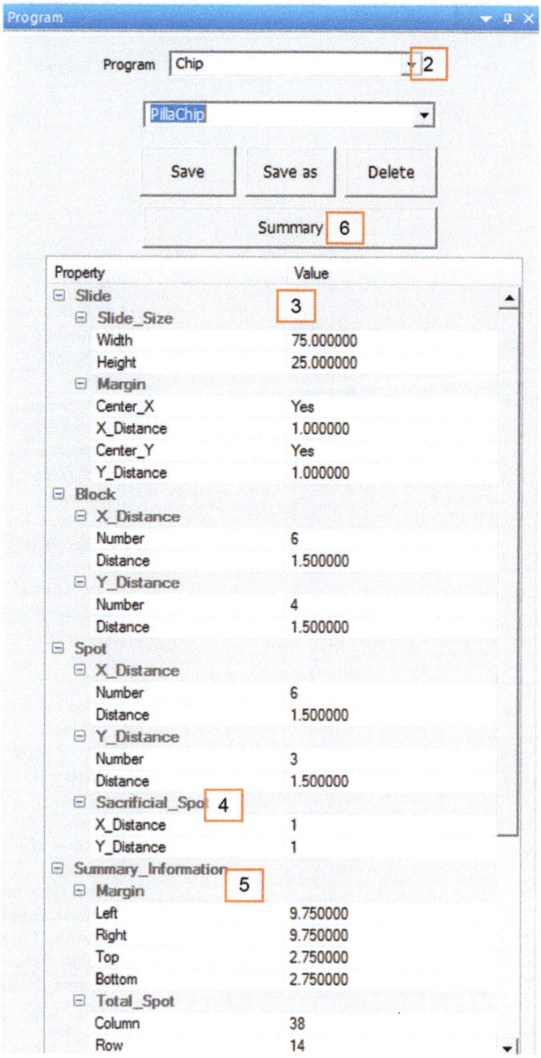

(A)

17. Click 'Save as' button to register the chip.
18. Enter a file name. ***Note: There is no space and special character allowed in the file name.***
19. If you want to modify the chip information you registered, select the file name in the box above the 'Save as' button and repeat Steps 3, 4, 15, and 16.
20. Click the 'Save' button to overwrite the changes in the same file.

Fig. 2.19 (continued)

(B)

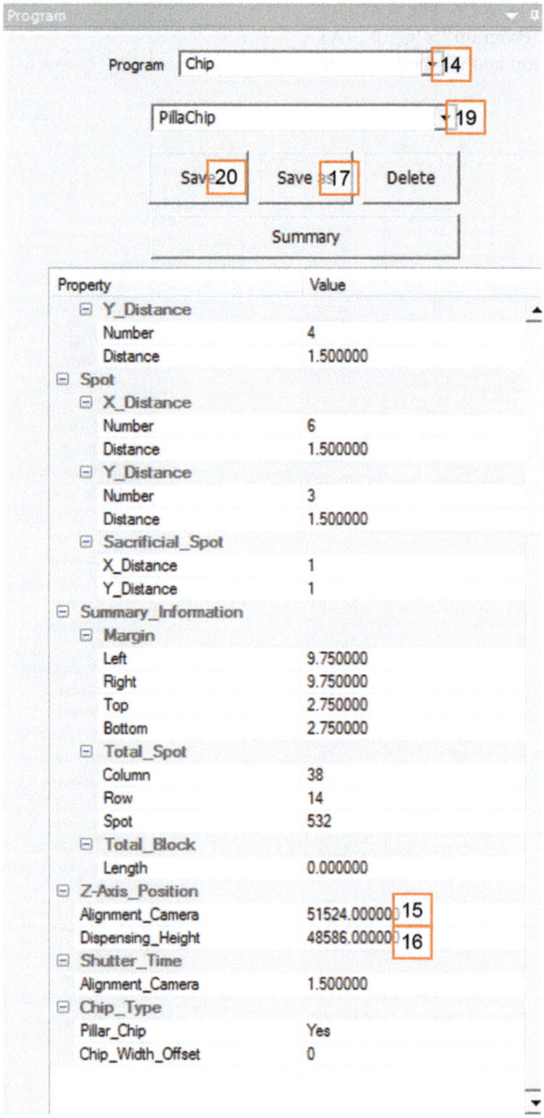

2.6.4 Registering Spot Layouts

1. Select the 'Program' window.
2. Select 'Spot Layout' in the 'Program' box.
3. Select the name of the chip you registered in the 'Slide/Chip' box (Fig. 2.22).
4. Check the chip information such as a margin, spot numbers, and a nozzle-to-nozzle distance.

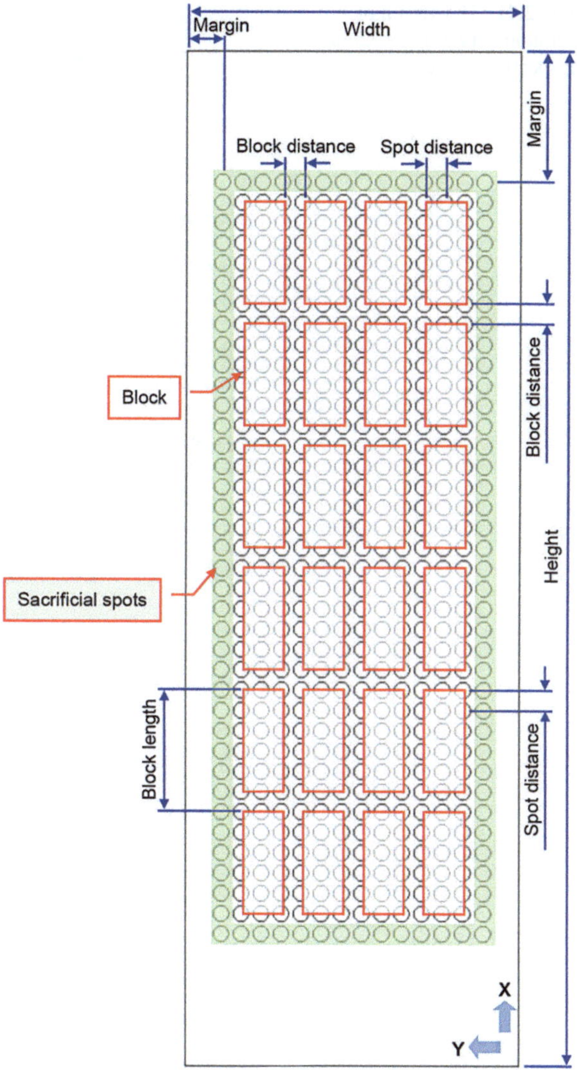

Fig. 2.20 The layout of the micropillar/microwell chip. The white circles indicate either micropillars or microwells on the chip platform.

5. Click '1st Spotting Layout' button to generate the first spotting pattern on the chip with samples.
6. Select 'Solenoid' in the 'Select Nozzle' box as by default we use it for sample spotting (Fig. 2.23).
7. Select the number of solenoid valves/ceramic tips and the layout used in the 'Nozzle Layout' box.
8. Select the type of the well plate used (96 or 384).
9. Select a 96-well (or a 384-well) where the first solenoid valve/ceramic tip will be located for sample aspiration by clicking the well. According to the number/layout of solenoid valves used, other wells will be automatically selected.

Fig. 2.21 The screen of 'Equipment Setup' > 'Chip Inspection Camera'.

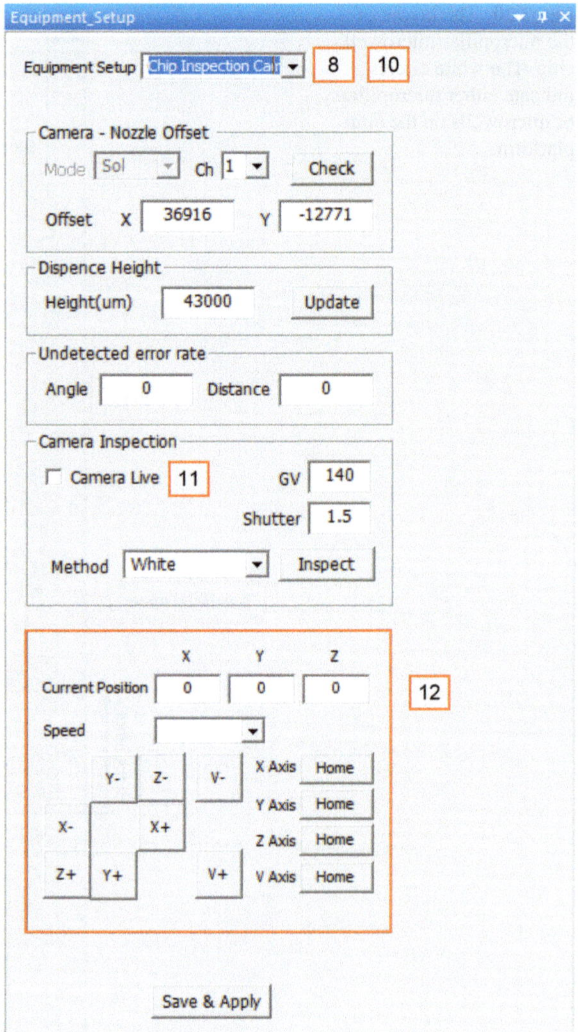

Note: For safety reasons, the entire well will not be selected when any one of the nozzles is left outside of the well plate for aspiration.

10. Select the region of spots where the sample will be dispensed with the first solenoid valve/ceramic tip by clicking and dragging on the chip layout.
11. Repeat Steps 9 and 10 according to the spot layout you want to generate.
12. Select the box beside 'Set Sacrificial Spot' to designate a sample well for printing sacrificial regions. ***Note: Always the first solenoid valve/ceramic tip is used for spotting sacrificial regions.***
13. Select the sample well for spotting sacrificial regions.
14. Click 'Close' button when done.
15. Click 'Save as' button to register the spot layout.

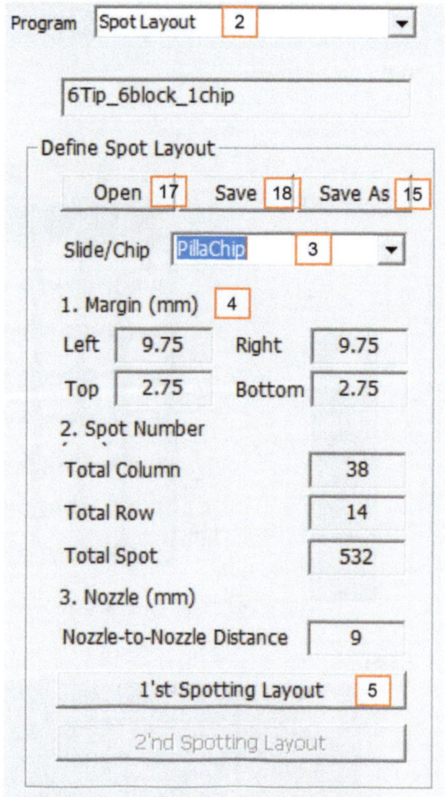

Fig. 2.22 The screen of 'Program' > 'Spot Layout'

16. Enter a file name. ***Note: There is no space and special character allowed in the file name.***
17. If you want to modify the spot layout you registered, click the 'Open' button and select the file name and repeat Steps 3 through 14.
18. Click the 'Save' button to overwrite the changes in the same file.

2.6.5 Optimizing Dispensing Parameters Using Vision Inspection

1. Select the 'User Operation' window.
2. Select 'Volume Check Vision' in the 'Daily Operation' box (Fig. 2.24).
3. Select the number of nozzles you want to inspect with the camera in the 'Nozzle' box.
4. Select the well plate used and indicate the location of a 96 or 384 well containing a test sample in the 'Sample Location' box. The first solenoid nozzle will be located in the well indicated, and the locations of the other solenoid

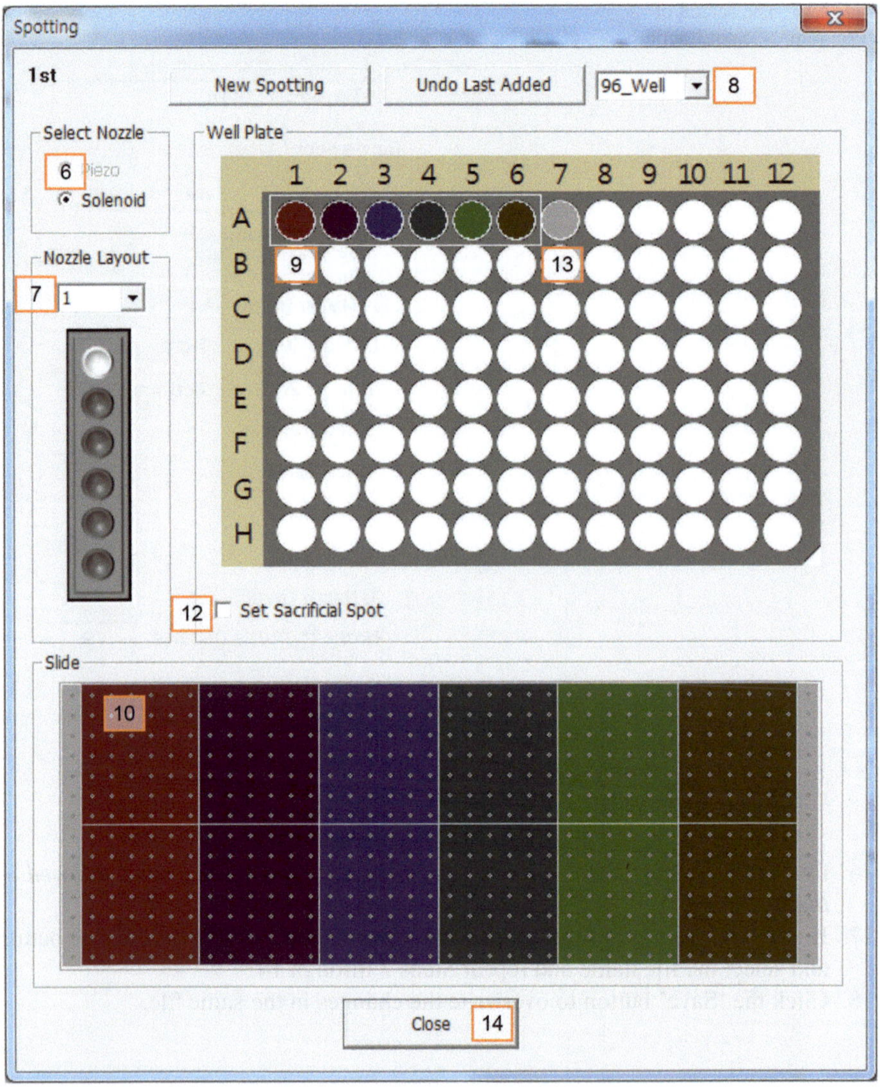

Fig. 2.23 The screen of 'Program' > 'Spot Layout'> '1st Spotting Layout'

valves/ceramic tips will be determined automatically according to the solenoid valves selected.

5. Enter the air pressure used for sample spotting and the volume of the sample loaded for inspection in the 'Pressure & Sample Aspiration Volume' box. The typical air pressure used is 6 kPa for 60–950 nL droplet.
6. Enter aspiration conditions such as air gap and pre-pressurization in 'Aspiration Condition Parameter'. Refer to Table 2.1.

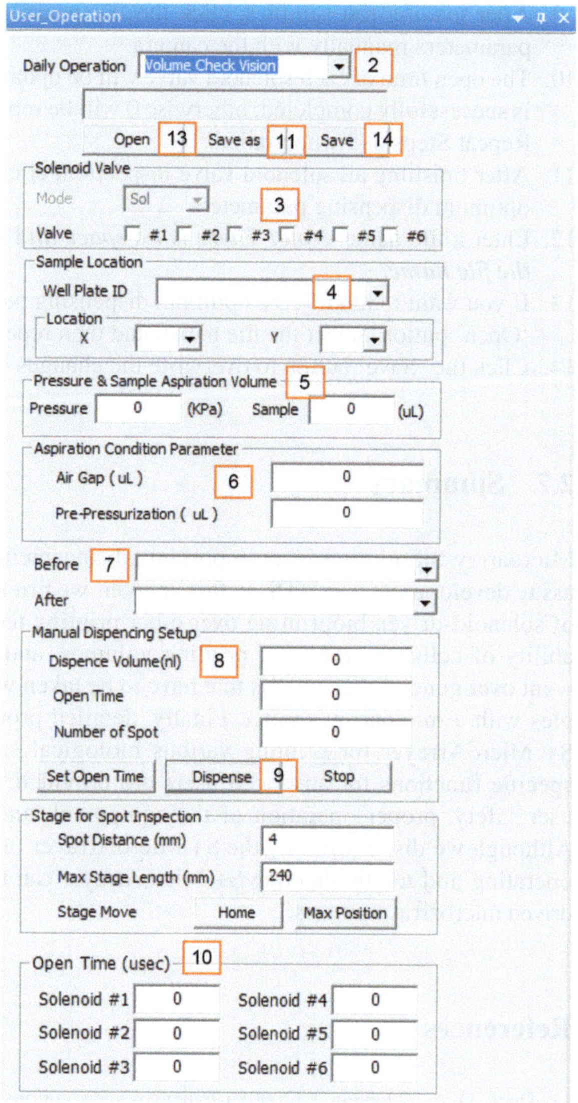

Fig. 2.24 The screen of 'User Operation' > 'Vision Inspection'

7. Select tip rinsing conditions in water before and after sample loading prior to sample dispensing in the 'Before' and 'After' boxes.
8. Enter the droplet volume for inspection in 'Dispensing Volume', the open time of the solenoid valves selected in 'Open Time', and the number of spots dispensed to test the conditions in 'Number of Spot' in the 'Manual Dispensing Setup' box. ***Note: If the disparity between the droplet volume set up and the average volume of the measured droplets at a certain condition is less than the desired CV value (typically 5 %), then the vision inspection will be successfully finished.***

9. After loading test samples, click the 'Dispense' button to test the dispensing parameters manually with the camera.
10. The open time of each solenoid valve will be updated when the vision inspection is successfully completed; otherwise 0 will be reported in the 'Open Time' box. Repeat Steps 3–9 when failed.
11. After finishing all solenoid valve inspection, click 'Save as' button to save the optimum dispensing parameters.
12. Enter a file name. ***Note: There is no space and special character allowed in the file name.***
13. If you want to modify the optimum dispensing parameters you saved, click the 'Open' button, select the file name, and then repeat Steps 3 through 11.
14. Click the 'Save' button to overwrite the changes in the same file.

2.7 Summary

Microarray bioprinting is an important advancement in the field of miniaturized assay development and HTS. In this chapter, we briefly introduced the advantages of solenoid-driven bioprinting over other printing technologies in terms of printability of cells, the range of printing volumes, and printing precision. We also went over general precautions that have to be taken when printing biological samples with a microarray spotter. Finally, detailed protocols on how to operate the S+ MicroArrayer for printing various biological samples and how to program specific functions for advanced users are provided. These protocols will ensure user safety, proper operation of the equipment, and consistent printing results. Although we discussed only the S+ MicroArrayer in this chapter, the principle of operating and troubleshooting S+ MicroArrayer can be applicable to any solenoid-driven microarray spotters.

References

1. Dunn, D. A., & Feygin, I. (2000). Challenges and solutions to ultra-high-throughput screening assay miniaturization: Submicroliter fluid handling. *Drug Discovery Today, 5*(12), 84–91. doi:10.1016/S1359-6446(00)80089-6.
2. Kapur, R., Giuliano, K. A., Campana, M., Adams, T., Olson, K., Jung., D., et al. (1999). Streamlining the drug discovery process by integrating miniaturization, high throughput screening, high content screening, and automation on the CellChip TM system. *Biomedical Microdevices, 2*(2), 99–109.
3. Horman, S. R., To, J., Orth, A. P., & Caracino, D. (2013). High-content analysis of three-dimensional tumor spheroids: Investigating signaling pathways using small hairpin RNA. *Nature Methods, 10*(10), v–vi. doi:10.1038/nmeth.f.370.
4. Justice, B. A., Badr, N. A., & Felder, R. A. (2009). 3D cell culture opens new dimensions in cell-based assays. *Drug Discovery Today, 14*(1), 102–107. doi:10.1016/j.drudis.2008.11.006.
5. Comley, J. (2013). Progress made in applying 3D cell culture technologies. *Drug Discovery World., 15*, 41–58.

6. Datar, A., Joshi, P., & Lee, M.-Y. (2015). Biocompatible hydrogels for microarray cell printing and encapsulation. *Biosensors, 5*(4), 647–663. doi:10.3390/bios5040647.
7. Salaita, K., Wang, Y., & Mirkin, C. A. (2007). Applications of dip-pen nanolithography. *Nature Nanotechnology, 2*(3), 145–155. doi:10.1038/nnano.2007.39.
8. Cho, E. J., & Bright, F. V. (2002). Integrated chemical sensor array platform based on a light emitting diode, xerogel-derived sensor elements, and high-speed pin printing. *Analytica Chimica Acta, 470*(1), 101–110. doi:10.1016/S0003-2670(02)00303-3.
9. Mota, C., & Moroni, L. (2015). Chapter 11—High throughput screening with biofabrication platforms. In A. Atala & J. J. Yoo (Eds.), *Essentials of 3D biofabrication and translation* (pp. 187–213). Amsterdam: Elsevier. doi:10.1016/B978-0-12-800972-7.00011-6.
10. Dusseiller, M. R., Schlaepfer, D., Koch, M., Kroschewski, R., & Textor, M. (2005). An inverted microcontact printing method on topographically structured polystyrene chips for arrayed micro-3-D culturing of single cells. *Biomaterials, 26*(29), 5917–5925. doi:10.1016/j.biomaterials.2005.02.032.
11. Zhang, L. G., Fisher, J. P., & Leong, K. W. (2015). *3D Bioprinting and Nanotechnology in Tissue Engineering and Regenerative Medicine*. Amsterdam: Elsevier. doi:10.1016/B978-0-12-800547-7.00004-7.
12. Boland, T., Xu, T., Damon, B., & Cui, X. (2006). Application of inkjet printing to tissue engineering. *The Biochemical Journal, 1*(9), 910–917. doi:10.1002/biot.200600081.
13. Cui, X., Boland, T., D'Lima, D. D., & Lotz, M. K. (2012). Thermal inkjet printing in tissue engineering and regenerative medicine. *Recent Patents on Drug Delivery & Formulation, 6*(2), 1–13. doi:10.2174/187221112800672949.
14. Nishiyama, Y., Nakamura, M., Henmi, C., Yamaguchi, K., Mochizuki, S., Nakagawa, H., et al. (2009). Development of a three-dimensional bioprinter: construction of cell supporting structures using hydrogel and state-of-the-art inkjet technology. *Journal of Biomechanical Engineering, 131*(3), 1–6. doi:10.1115/1.3002759.
15. PK, W., Ringeisen, B. R., Callahan, J., Brooks, M., Bubb, D. M., Wu, H. D., et al. (2001). The deposition, structure, pattern deposition, and activity of biomaterial thin-films by matrix-assisted pulsed-laser evaporation (MAPLE) and MAPLE direct write. *Thin Solid Films, 398–399*, 607–614. doi:10.1016/S0040-6090(01)01347-5.
16. Devillard, R., Pagès, E., Correa, M. M., Kériquel, V., Rémy, M., Kalisky, J., et al. (2014). Cell patterning by laser-assisted bioprinting. *Methods in Cell Biology, 119*, 159–174. doi:10.1016/B978-0-12-416742-1.00009-3.
17. Catros, S., Guillotin, B., Bačáková, M., Fricain, J. C., & Guillemot, F. (2011). Effect of laser energy, substrate film thickness and bioink viscosity on viability of endothelial cells printed by Laser-Assisted Bioprinting. *Applied Surface Science, 257*(12), 5142–5147. doi:10.1016/j.apsusc.2010.11.049.
18. Demirci, U., & Montesano, G. (2007). Single cell epitaxy by acoustic picolitre droplets. *Lab on a Chip, 7*(9), 1139–1145. doi:10.1039/b704965j.
19. Fang, Y., Frampton, J. P., Raghavan, S., Sabahi-Kaviani, R., Luker, G., Deng, C. X., et al. (2012). Rapid generation of multiplexed cell cocultures using acoustic droplet ejection followed by aqueous two-phase exclusion patterning. *Tissue Engineering. Part C, Methods, 18*(9), 647–657. doi:10.1089/ten.tec.2011.0709.
20. Demirci, U., & Montesano, G. (2007). Cell encapsulating droplet vitrification. *Lab on a Chip, 7*(11), 1428–1433. doi:10.1039/b705809h.

Chapter 3
Chip Platforms and Chip Surface Treatments

Parnian Bigdelou, Alexander Roth, Yana Sichkar, and Moo-Yeal Lee

Contents

3.1 Introduction 54
 3.1.1 Surface Modification of Micropillar and Microwell Chips 54
 3.1.2 Surface Modification of Glass Slides 56
3.2 Materials 57
3.3 Protocols 58
 3.3.1 Cleaning Microwell Chips with Plasma for Air Bubble Removal 58
 3.3.2 Coating Micropillar Chips with Poly(maleic anhydride-*alt*-1-octadecene) (PMA-OD) for Cell Printing 61
 3.3.3 Cleaning the Surface of Glass Slides with Acid 61
 3.3.4 Coating of Glass Slides with Poly(styrene-*co*-maleic anhydride) (PS-MA) for Cell Printing 63
 3.3.5 Coating of Acid-Cleaned Glass Slides with Methyltrimethoxysilane (MTMOS) for Enzyme Printing 64
 3.3.6 Coating of Acid-Cleaned Glass Slides with (3-Aminopropyl)trimethoxysilane (APTMS) for Protein Attachment 65
 3.3.7 Measurement of Silanization on the Surface of the APTMS Slides with Fluorescein Isothiocyanate (FITC) Labeling 65
 3.3.8 Attachment of Glutaraldehyde on the Surface of the APTMS Slides for Protein Attachment 66
 3.3.9 Immobilization of Extracellular Matrix (ECM) Proteins on Reactive Surfaces 67
3.4 Summary 68
References 69

M.-Y. Lee, *Microarray Bioprinting Technology*,
DOI 10.1007/978-3-319-46805-1_3

3.1 Introduction

There are three significant types of platforms used in microarray printing technology: glass slides, micropillar chips, and microwell chips. The surfaces of these platforms are treated for different applications. For example, a glass slide is ideal for micropatterning and can be used for microscopic image analysis. Micropillar and microwell chips are ideally suited for applications of high-throughput screening (HTS) of compounds on 3D-cultured cells. In this chapter, we present various methods by which these surfaces can be chemically modified for use in bioprinting applications, with emphasis on the methods used for culture of cells encapsulated in hydrogels.

3.1.1 Surface Modification of Micropillar and Microwell Chips

Since many cells have difficulty in attaching to the surface of glass slides, different coatings on glass slides have been developed to improve cell attachment and growth, such as using rat tail collagen. In addition, glass slides need very tedious and precise cleaning procedures to make sure that any residues on the surface do not affect cellular processes. To alleviate these issues and facilitate robust and reproducible cell cultures, our research team developed disposable plastic chip platforms such as a micropillar chip and a microwell chip (Fig. 3.1).

Both micropillar and microwell chips are made of polystyrene, a long carbon chain polymer with benzene rings attached to every other carbon (Fig. 3.2). Polystyrene has a very good optical clarity, is easily moldable, and can be sterilized by UV or plasma irradiation. However, it is hydrophobic in nature without functional groups; thus, cells and hydrogels with cells have difficulty in attaching on its surface. In order to have proper attachment, the surface of polystyrene is modified

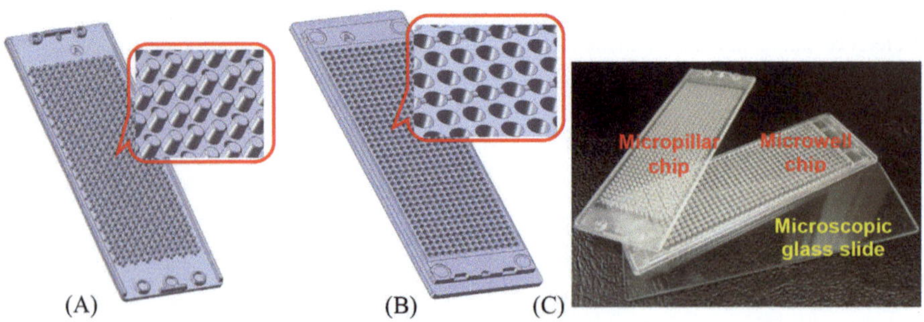

Fig. 3.1 Schematics of (**A**) a micropillar chip and (**B**) a microwell chip. (**C**) Image of micropillar and microwell chips to scale against a microscopic glass slide

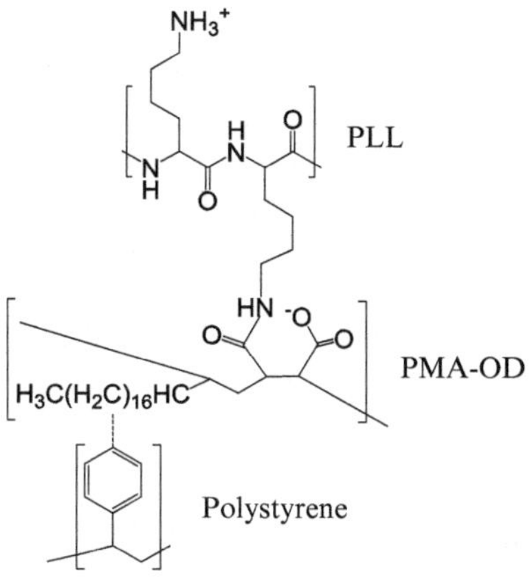

Fig. 3.2 Modification of polystyrene surface by coating with poly(maleic anhydride-*alt*-1-octadecene) (PMA-OD) and subsequent covalent attachment of poly-L-lysine (PLL)

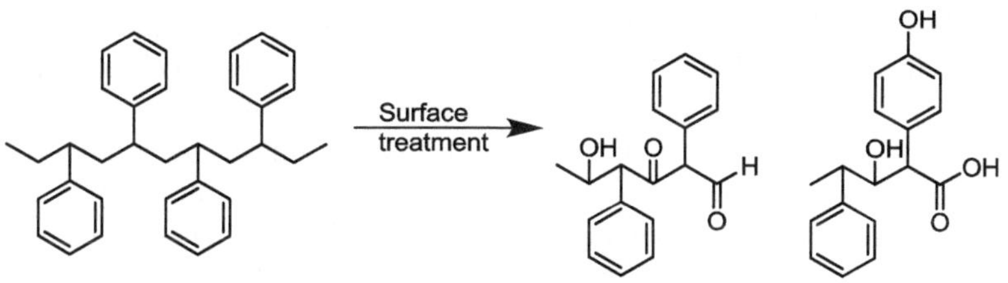

Fig. 3.3 Modification of polystyrene surface via plasma treatment. A variety of different chemical groups can be introduced by breaking the carbon chain backbone

to allow strong interactions between cell surface/hydrogel and polystyrene surface. For example, cells can be attached more easily on negatively charged or amine-reactive surfaces in general, because extracellular matrices (ECMs) secreted by the cells as well as serum proteins in culture media can be attached easily on these surfaces, therefore providing a better surface for cell attachment.

One efficient method for surface modification of polystyrene involves using amphiphilic co-polymers, utilizing the hydrophobic interactions of the polymer with the polystyrene, while functional side chains can provide an anchor for hydrogels or ECMs to attach. One such polymer, poly(maleic anhydride-*alt*-1-octadecene) (PMA-OD) has been used for improved hydrogel attachment to micropillar chips. When the cells are encapsulated in hydrogels, such as alginate, they can attach to the surface of the polystyrene chip, through the interactions of PMA-OD and the amine-reactive residues on the hydrogel. Subsequent improvements to gel adhesion can be made by using poly-L-lysine (PLL), utilizing the amine groups on PLL as an anchoring point for hydrogel attachment, as shown in Fig. 3.2. The positively charged PLL also interacts with negatively charged hydrogels and supports the attachment of hydrogels on the micropillar chip [1].

The hydrophobic surface of polystyrene can cause an additional problem such as an entrapment of air bubbles in the microwells. One effective method to convert the hydrophobic surface of polystyrene to hydrophilic is to irradiate the surface with strong gas plasma under vacuum. This process generates highly energetic oxygen ions and radicals, which oxidizes the polystyrene chains (Fig. 3.3) so that the surface becomes hydrophilic and negatively charged [2–5]. This process uses an ionized gas or plasma, to remove all organic matter from the surface of an object, which is carried out in a vacuum chamber. During plasma treatment, excited gas molecules and atoms in the plasma will emit UV light. By introducing different gases into the plasma chamber such as O_2, N_2, Ar, He, and CF_4, treatment can be performed. There are no harsh chemicals involved in the process of plasma cleaning; therefore, it is an environmentally safe process.

3.1.2 Surface Modification of Glass Slides

Glass slides have significant use for biological testing. The distinct advantages of glass substrates are that they are relatively non-reactive, which is suitable for bio-compatibility, and they are well suited for imaging purposes. Additionally, glass slides are thermally stable, which makes them relatively easy to clean and modify for future uses. The main struggle with glass surfaces is that their non-reactivity, while potentially useful for culturing cell monolayers, makes them a poor surface for hydrogel attachment. This hydrogel attachment is critical, as hydrogels and other structures have the potential for cells to organize themselves into three-dimensional (3D) structures, mimicking the cell-ECM interactions found *in vivo*. If hydrogels are not properly attached, any treatment towards the glass slide can result in subsequent loss of cells encapsulated in hydrogels. Thus, it is a goal of researchers to investigate methods for robustly immobilizing hydrogels onto glass slides without compromising the integrity of the hydrogel or the normal behavior of the cells.

Silicone precursor molecules are important to initiate this attachment between glass slides and biological molecules. The strong covalent bonding between the glass made of silicone oxide and the silicone precursor allows for the interaction of biologics (such as hydrogels, proteins, and cell surfaces) with the functionalized glass slide with the silicone precursor. One potential method that has been used to improve protein and hydrogel attachment onto glass slides is the use of silanes with amine groups, which can covalently attach biological compounds to silane surfaces. The silicon in the silane group anchors well with the silicon glass slide, while functional groups attached to the silane can interact with functional groups on hydrogels and cell membranes to promote robust attachment. Such functional groups include thiols, esters, N-hydroxysuccinimide (NHS)-esters, amines, and methacrylates [6]. Seo *et al.* demonstrated the use of thiol silanes and acryl silanes for the covalent attachment of polyethylene glycol (PEG) hydrogels and streptavidin onto glass slides without compromising the hydrogel integrity or the affinity of the streptavidin-biotin interaction [6]. While silanes with initial functional groups can be maintained on a glass surface, H-terminal silanes can also be modified. Carvalho *et al.* demonstrated the conjugation of alkenes with H-terminal silanes on glass via ultraviolet (UV) light [7]. Luo *et al.* used a similar procedure, but applied acrylamide onto the surface for a different surface chemistry [8]. The use of UV light is important for treating glass surfaces, as it serves the dual function of modification of the surface, and sterilization.

While UV light has been used as a mechanism for chemical modification of glass surfaces, the thermal stability of silicon makes glass a suitable surface for thermally controlled surface modification. Cheawchan *et al.* has demonstrated this by initially modifying the glass surface with allyl groups before 70 °C heating was applied, and the surface was reacted with nitrile n-oxide moieties [9]. The glass slide can then be further modified for addition of functional groups that can aid in the attachment of cells and hydrogels to the glass surface.

Another potential method for improving the attachment of biological molecules onto glass slides involves initial surface treatments with gold. Gold can be easily coated onto the surface of glass slides using various physical methods, but vapor deposition is among the most common ways for entirely plating gold onto a glass slide [10]. This is followed up with subsequent attachment of thiol groups [10]. The covalent attachment of sulfur on gold is easily achievable. This means that cysteine and potentially methionine residues can be targeted on proteins for immobilization purposes. Additionally, disulfide linkages can be used for a potential point of interaction between gold and any thiol residues on the surface of the glass slide.

3.2 Materials

- Micropillar/microwell chips (25 × 75 × 2 mm from SEMCO, Suwon, South Korea)
- Fisherbrand plain microscope slides (25 × 75 × 1 mm from Thermo Fisher Scientific, Rockford, IL) or custom precision glass slides (25.35 × 75.50 × 1.10 mm from Coresix Precision Glass, Inc., Williamsburg, VA)
- Wheaton glass 20 slide staining dish with removable rack (Fisher)
- National scientific 20 mL glass sample vials with PTFE-lined solid storage caps (Fisher)
- Methyltrimethoxysilane (MTMOS from Sigma-Aldrich, St. Louis, MO)
- 3-(Aminopropyl)trimethoxysilane (APTMS from Sigma-Aldrich)
- Poly(maleic anhydride-*alt*-1-octadecene) (PMA-OD from Sigma Aldrich)
- Poly(styrene-*co*-maleic anhydride), 14 % (w/w) maleic anhydride content (PS-MA from Sigma-Aldrich)
- Glutaraldehyde
- Matrigel™, collagen I, poly-L-lysine, fibronectin, laminin, vitronectin, *etc.*
- Fluorescein isothiocyanate (FITC)
- Dulbecco's phosphate-buffered saline (PBS) without Ca^{2+} and Mg^{2+}
- Potassium phosphate buffer solution (200 mM, pH 8 from Life Technologies, Carlsbad, CA).
- Nalgen MF75 series sterile disposable tissue culture filter unit (0.2 µm PES filter)
- Spin coater (Model PWM32, Headway Research, Inc., Garland, TX)
- Plasma cleaner (Harrick Plasma, PDC-001-HP)

- Vacuum pump (Dekker Vacuum Technologies, RVR003H-01)
- Digital thermocouple gauge control unit (Duniway Stockroom Corp, DTC531S-115-BX)
- S+ MicroArrayer with six solenoid valves and ceramic tips (SEMCO, Suwon, South Korea)
- S+ Scanner with four filter sets (SEMCO, Suwon, South Korea)

3.3 Protocols

3.3.1 Cleaning Microwell Chips with Plasma for Air Bubble Removal

Plasma treatment of microwell chips is carried out to change the surface property of the microwell chip hydrophilic and remove air bubbles trapped in the microwells when immersed in growth media or to sterilize the microwell chips. The protocol of plasma treatment provided here is modified from the user manual from the manufacturer [11].

1. Make sure the machine and the thermocouple are plugged into an appropriate power source, and make sure the vacuum pump is plugged into the plasma cleaner. Turn on the power switch located on the vacuum pump, but do not touch the 'PUMP' switch on the Plasma Cleaner's panel just yet.
2. Turn on the Plasma Cleaner's main power ('POWER' switch to ON).
3. Open the door of the Plasma Cleaner. Load microwell chips on the Pyrex glass tray and put it inside the Plasma Cleaner's chamber. Maximum ten chips can fit on the tray (Fig. 3.4). ***Note: Place the chips facing up in the position for***

Fig. 3.4 The position of the microwell chips on the Pyrex glass tray

plasma treatment. In addition, wipe the back of the microwell chips with 70 % ethanol-soaked paper towels to prevent microbial contamination.

4. Check that the 3-way valve is closed. It must be leveled in the vertical downward position (Fig. 3.5).
5. Close and lean the door tightly against the Plasma Cleaner.
6. While holding the door, set the Plasma Cleaner's 'PUMP' switch to ON. It is located directly on the Plasma Cleaner (Fig. 3.6). The vacuum will maintain the door tightly closed. Wait for 5 min.
7. Level the 3-way valve into the horizontal position such that it points to the left (Fig. 3.7).
8. Slightly open the metering valve and wait for 1 min (Fig. 3.7). It will allow the air to enter the Plasma Cleaner. ***Note: The metering valve should be operated in the way the plasma intensity is noticeably increased and the pressure at the thermocouple is ranging from 800 to 1000 mTorr. Other pressures may also work depending on the samples, but 900–1000 mTorr is ideal for microwell chips.***
9. Turn on Plasma Cleaner's RF switch to 'HI'. Wait for 1 min. A fluorescence should be observed through the Plasma Cleaner's window. For a plasma cleaning process with room air, purple-pink to violet plasma should be observed (Fig. 3.8).

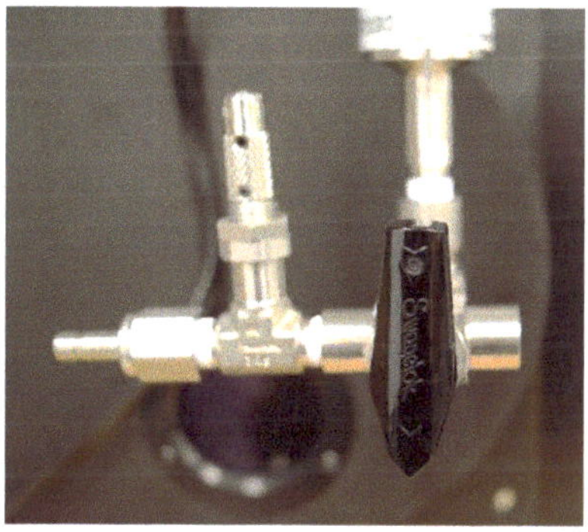

Fig. 3.5 Vertical downward position of the 3-way valve. The 3-way valve is closed in this position

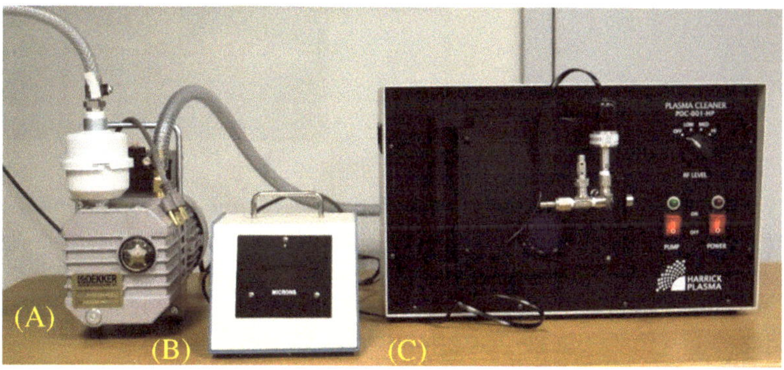

Fig. 3.6 Plasma treatment of microwell chips is performed using (**A**) a vacuum pump, (**B**) a digital thermocouple gauge control unit, and (**C**) a plasma cleaner unit

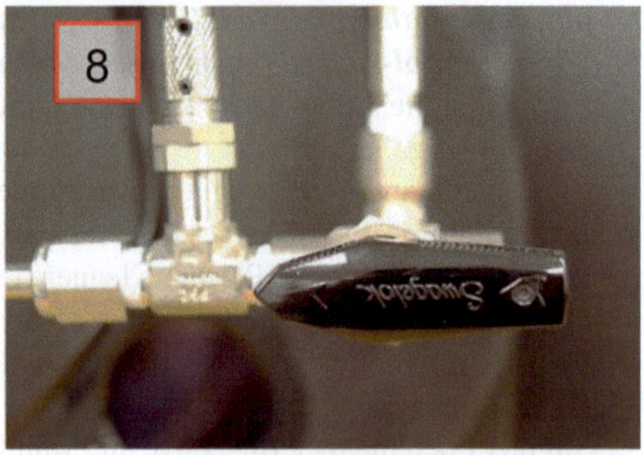

Fig. 3.7 The 3-way valve is positioned in the *horizontal* position such that it points to the *left*

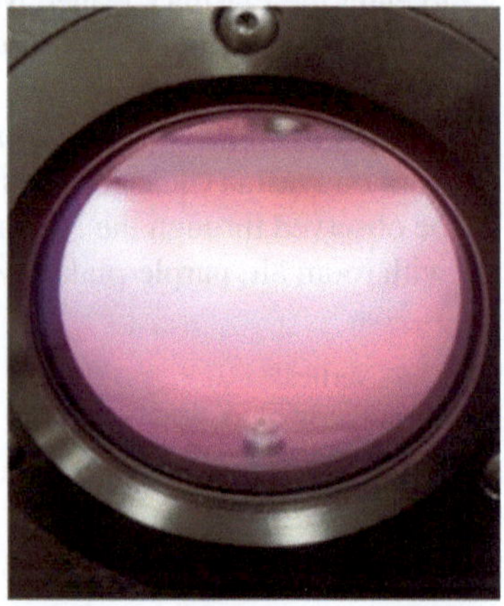

Fig. 3.8 *Purple-pink* plasma generated during the plasma treatment of microwell chips

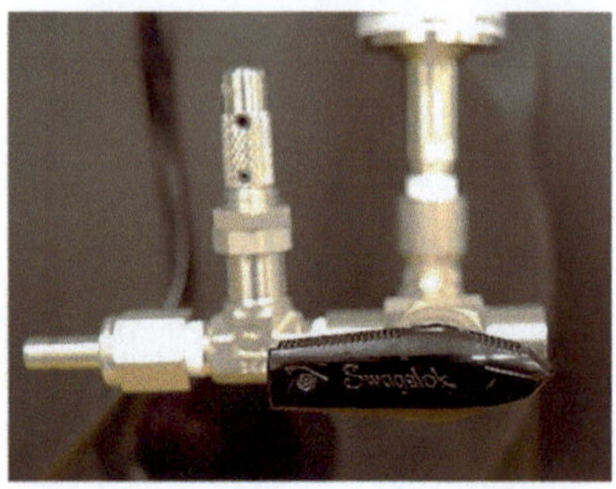

Fig. 3.9 The 3-way valve is positioned in the *horizontal* position such that it points to the *right* which allow the system to vent

10. Process samples for 15 min. ***Note: The process time has to be optimized for different applications.***
11. At the end of the process, switch the RF level to 'OFF'.
12. Level the 3-way valve to the vertical downward position (Fig. 3.5)
13. Turn off the vacuum pump by setting the Plasma Cleaner's 'PUMP' switch to OFF.
14. **Slowly** level the 3-way valve into the horizontal position such as it points to the right (Fig. 3.9). The system will vent. ***Note: The Plasma Cleaner must be vented***

instantly. Otherwise, oil from the vacuum pump may back stream from the pump and contaminate the system. Do not open the front door when the chamber is under vacuum since this will damage the glass sample holder and any samples. Additionally, a slow vent ensures that the samples and the Pyrex tray do not move around the chamber when the samples are removed.

15. When the hissing noise from venting the chamber is not observed anymore, open the front door.
16. Close the 3-way valve. Level it to the vertical downward position (Fig. 3.5).
17. Turn off the Plasma Cleaner's main power ('POWER' switch to OFF).
18. Remove the chip samples.
19. Unplug the thermocouple at the end of the process, and keep it unplugged when it is not used.

3.3.2 Coating Micropillar Chips with Poly(maleic anhydride-alt-1-octadecene) (PMA-OD) for Cell Printing

The amphiphilic PMA-OD coating is necessary to create an amine-reactive and hydrophobic surface of the micropillar chip for cell printing.

1. Prepare a 1 % (weight/volume or w/v) PMA-OD in ethanol solution by dissolving 200 mg of PMA-OD in 20 mL of 100 % ethanol and letting it mix with a small stir bar in a sealed glass sample vials with PTFE-lined solid storage caps for at least 2 days for dissolution.
2. Dilute slightly turbid 1 % PMA-OD with 100 % ethanol at 1:10 ratio and mix properly to prepare a clear 0.1 % PMA-OD stock solution.
3. Further dilute 0.1 % PMA-OD with 100 % ethanol at 1:10 ratio and mix properly to prepare a 0.01 % PMA-OD working solution.
4. Add 2 mL of 0.01 % PMA-OD on the shallow staining plate and sandwich the micropillar chip (Fig. 3.10).
5. Pick up and slant the micropillar chip at an angle of 45° for 1 min to allow excess PMA-OD on the surface to drain, and wipe off the solution at the bottom of the chip with Kimwipes.
6. Blow dry the bottom surface of the micropillar chips, then let them dry at room temperature with the micropillars facing up in a sterile bioassay plate (Fig. 3.11).

3.3.3 Cleaning the Surface of Glass Slides with Acid

For a very even silanization of a glass surface as well as uniform polymer coating, removing all dirt and exposing silanol group (—SiOH) on the surface of glass slides with strong acid are extremely important [12]. Unless pre-cleaned precision glass

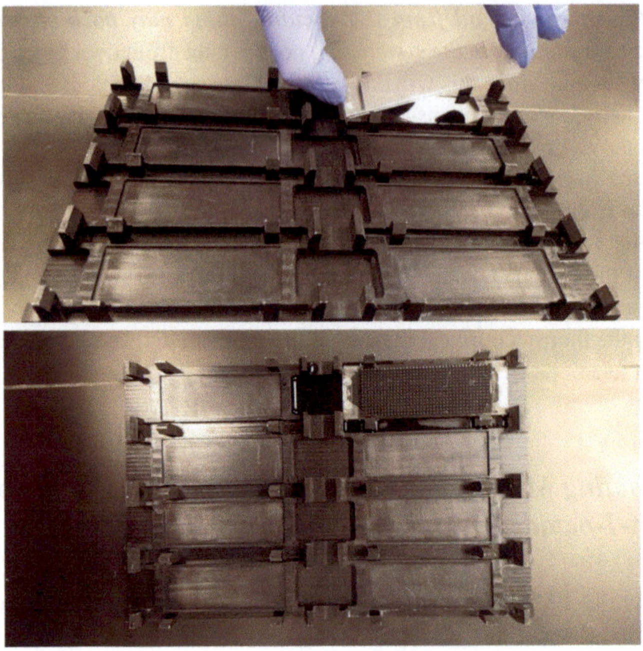

Fig. 3.10 Coating the surface of a micropillar chip with 0.01 % (w/v) PMA-OD in a shallow staining plate

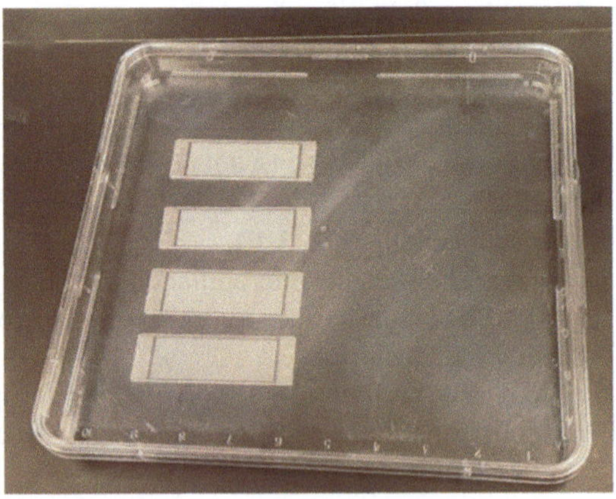

Fig. 3.11 Micropillar chips facing up in a sterile bioassay plate

slides from Coresix Precision Glass, Inc. are used, the acid-cleaning step is essential for standard microscope slides.

1. Properly number plain microscope glass slides (25 × 75 × 1 mm) with a diamond tip.
2. Wipe the glass slides three times with 70 % ethanol-soaked paper towels to remove oil and solid particles on the slide surface and then clean ethanol with

dry paper towels. ***Note: Do not use Kimwipes because it leaves lots of small paper particles.*** After wiping with dry paper towels, the microscope slides should not be wet with ethanol. It will leave ethanol stains on the slide surface.
3. Place the glass slides in a removable glass rack, immerse the rack in a Wheaton glass dish filled with concentrated sulfuric acid (96.5 %) for 2 h, and then sonicate the Wheaton glass dish with the slides in acid for 10 min. ***Note: Always work in a chemical fume hood while handling acid or organic solvent, do not inhale acidic fume, and do not reuse sulfuric acid more than twice.*** Dispose of all chemicals in a labeled waste container properly.
4. Rinse the rack containing the glass slides at least five times with de-ionized distilled water to remove excess sulfuric acid on the slide surface.
5. Immerse the rack containing the slides twice in a Wheaton glass dish filled with acetone to remove excess water on the glass slide and dry the acid-cleaned glass slides thoroughly under N_2 gas stream.
6. Bake the acid-cleaned slides in an oven at 120 °C for 10 min to completely remove water on the slide surface.
7. After drying, carefully inspect the glass slide for cleanliness. If the slide is not clean enough (i.e., any dirt or particle on the slide surface), return to the Step 2. ***Note: Always use tweezers after the Step 2. Do not use fingers to hold the slides, as this will leave an oily residue on the slide.***
8. Store the glass slides in a sterile petri dish (150 mm diameter, maximum 5 slides/ petri dish) or the bioassay plate until use. Never leave the clean slide uncovered for long time.

3.3.4 Coating of Glass Slides with Poly(styrene-co-maleic anhydride) (PS-MA) for Cell Printing

To attach cell spots or prepare cell patterns on the surface of glass slides, it is required to modify the surface amine-reactive while enhancing hydrophobicity to prevent spot spreading. The amphiphilic PS-MA coating is suitable to change the hydrophilic surface of glass slides to hydrophobic while providing amine-reactive functionality for cell spot attachment.

1. Prepare a fresh 1 % (w/v) solution of PS-MA by dissolving PS-MA in anhydrous toluene in a scintillation vial with a PTFE-coated lid. ***Note: Use disposable glass serological pipets to aspirate toluene precisely.***
2. Shake the solution in an incubating shaker at 50 °C, 150 rpm for 40 min until PS-MA pellets completely dissolved. ***Note: After 30 min of shaking at 50 °C, 150 rpm, the solution usually become clear.*** However, leave the solution for additional 10 min for complete dissolution of PS-MA pellets. ***The PS-MA solution should be prepared freshly as reactivity of maleic anhydride groups decreases over time. Additional note: The PTFE-lined scintillation vial has to be used for organic solvents. Never leave the vial with toluene in the incubat-***

ing shaker unattended due to a concern about an explosion at high temperature. Always keep the maximum shaker temperature below 70 °C for safety precaution as the boiling point of toluene is 110.6 °C.
3. Prepare a 0.1 % (w/v) PS-MA solution by mixing 18 mL of anhydrous toluene with 2 mL of the 1 % (w/v) PS-MA solution. ***Note: Prepare a fresh PS-MA solution and use it immediately.***
4. Prior to spin coating, bake acid-cleaned slides or amine-functionalized slides in an oven at 120 °C for 10 min to completely remove moisture on the slide surface.
5. Load 1 mL of the 0.1 % (w/v) PS-MA solution onto the slide and spin coating the solution at 3000 rpm for 30 s (Model PWM32, Headway Research, Inc., Garland, TX).
6. Remove any excess PS-MA remaining on the back of the slide by wiping with acetone-soaked paper towel.
7. Dry the PS-MA-coated slides overnight at room temperature in a sterile petri dish and store until use.

3.3.5 Coating of Acid-Cleaned Glass Slides with Methyltrimethoxysilane (MTMOS) for Enzyme Printing

For enzyme printing, amine-reactive surfaces may deactivate the enzymes printed. Thus, the surface of acid-cleaned glass slides is typically changed to hydrophobic. A commonly used method is to coat the surface of the glass slides with hydrophobic silicone precursors such as MTMOS.

1. Prepare a fresh MTMOS-HCl sol solution by mixing 8 mL of MTMOS with 3.2 mL of 5 mM HCl, followed by vortex for 2 min and sonication for 10 min. ***Note: The hydrolysis of MTMOS by HCl is an exothermic reaction. So be careful not to burn skin.*** The mixture becomes transparent single phase after vortex mixing (initially two phases). Always prepare fresh MTMOS-HCl sol solution and use it immediately. Leaving the MTMOS-HCl sol solution longer than 30 min before spin coating cause uneven coating.
2. Immediately before spin coating, mix 11.2 mL of the MTMOS-HCl sol solution with 8 mL of potassium phosphate buffer solution (25 mM, pH 8) and use the mixture within 15 min. ***Note: Upon addition of the phosphate buffer solution, gelation of hydrolyzed MTMOS begins.*** Therefore, use the mixture immediately. Do not use the mixture when the solution becomes turbid, causing uneven coating.
3. While spinning the acid-cleaned glass slide at 3000 rpm for 30 s, place 1.5 mL of the mixture onto the slide.
4. Remove any excess MTMOS on the back of the slide using acetone-soaked paper towel. ***Note: Make sure that rainbow-like refraction patterns are seen on the surface after spin coating.*** If not, then the MTMOS coating is unsuccessful,

presumably because either the acid-cleaned glass slide is too dirty or the MTMOS-HCl-phosphate buffer mixture is too old.

5. To complete MTMOS gelation, dry the slides overnight at room temperature in a sterile petri dish and store until use.

3.3.6 Coating of Acid-Cleaned Glass Slides with (3-Aminopropyl)trimethoxysilane (APTMS) for Protein Attachment

Diluted APTMS solutions can be used to achieve amine functionalization on glass slides, which can be used for protein attachment using a linker.

1. Place acid-cleaned glass slides in a removable glass rack, and then immerse the rack in a Wheaton glass dish filled with acetone to completely remove moistures on the surface.
2. After acetone rinsing, dry the slides by blowing nitrogen.
3. Prepare 5 % (v/v) of APTMS in toluene containing 0.5 % (v/v) of methylene chloride (CH_2Cl_2). ***Note: Concentrated APTMS should not be exposed to air and humidity to prevent oxidation and hydrolysis.*** Make a fresh APTMS solution each time for surface functionalization [13]. Toluene is a carcinogenic substance. Thus, do not inhale the fume, and handle it in a chemical hood.
4. Immerse the glass slides in a fresh APTMS solution and sonicate them for 1 h. Discard the used APTMS solution after sonication.
5. Rinse the glass slides three times by dipping in toluene for 3 min and then remove excess toluene by dipping in acetone. ***Note: Do not dry the slides while rinsing. Drying can cause uneven silanization.*** Do not sonicate the slides while rinsing because sonication doesn't help to remove multiple layers of APTMS. Sonication just causes non-uniform amine density on the surface.
6. Dry acetone on the slides by blowing nitrogen
7. Bake the slides at 120 °C for 1 h.
8. After cooling to room temperature, rinse the slides in ethanol three times for 20 min to remove uncoupled APTMS.
9. After drying, store amine functionalized glass slides in a petri dish until use.

3.3.7 Measurement of Silanization on the Surface of the APTMS Slides with Fluorescein Isothiocyanate (FITC) Labeling

1. Allow DMSO (component B) and one vial of reactive FITC dye (component A) from Life Technologies (FluoReporter® protein labeling kit) to warm to room temperature immediately before starting the reaction.

2. Add 50 μL of DMSO to the vial of reactive FITC dye. ***Note: Protect the reactive FITC dye from light.***
3. Prepare 200 mL of 50 mM potassium phosphate buffer (pH 8 from Life Technologies) by diluting with de-ionized distilled water. ***Note: The buffer solution should not contain any ammonium ions or primary amines because the dye is reactive with amine groups. Thus, Tris or glycine buffers are unsuitable.***
4. Add 50 μL of reactive FITC dye in 200 mL of phosphate buffer and mix well with a magnetic stir bar. Any remaining reactive dye stock solution should be discarded.
5. Place amine-functionalized slides and acid-cleaned glass slides (as a control) in a removable glass rack, and then immerse the rack in a Wheaton glass dish filled with 100 % acetonitrile (ACN) for 5 min.
6. Rinse the slides in series for 5 min with an equal volume mixture of ACN and water, and de-ionized distilled water to wet the surface.
7. Immerse the slides in the reactive FITC dye solution and incubate them for 1 h with magnetic stirring at room temperature. ***Note: Do not sonicate the slides in the reactive FITC dye solution because sonication can cause non-specific binding of the dye on the surface.*** While incubating the slides, protect the fluorescent dye solution from light by wrapping it up in aluminum foil.
8. Rinse the slides three times in de-ionized distilled water for 1 h each with magnetic stirring to remove unbound dye.
9. Rinse the slides in acetone to remove excess water and wrap them up in aluminum foil. ***Note: Be careful not to scratch the surface of the slides with aluminum foil.***
10. Measure fluorescent intensity on the slides at 494 nm (excitation) and 518 nm (emission).

3.3.8 Attachment of Glutaraldehyde on the Surface of the APTMS Slides for Protein Attachment

1. Rinse the amine-functionalized slides in series for 5 min with 100 % acetonitrile (ACN), an equal volume mixture of ACN and water, and de-ionized distilled water to wet the surface. Sterilize de-ionized distilled water with Nalgen MF75 series sterile disposable tissue culture filter unit (0.2 μm PES filter).
2. Prepare 200 mL of 5 % (v/v) of glutaraldehyde in sterile Dulbecco's phosphate-buffered saline (PBS) without $CaCl_2$ & $MgCl_2$.
3. Immerse the amine-functionalized slides in the glutaraldehyde solution and sonicate them for 1 h at room temperature. Discard the used glutaraldehyde solution after incubation.
4. Rinse the slides twice by dipping in de-ionized distilled water for 5 min to remove unbound glutaraldehyde. Use sterile de-ionized distilled water.
5. Rinse the slides in acetone and dry them by blowing nitrogen. ***Note: Use the slides as soon as prepared because of instability of glutaraldehyde.***

3.3.9 *Immobilization of Extracellular Matrix (ECM) Proteins on Reactive Surfaces*

ECM proteins can be covalently attached on the reactive surfaces of glass slides such as PS-MA-coated slides and glutaraldehyde-functionalized slides (Figs. 3.12 and 3.13).

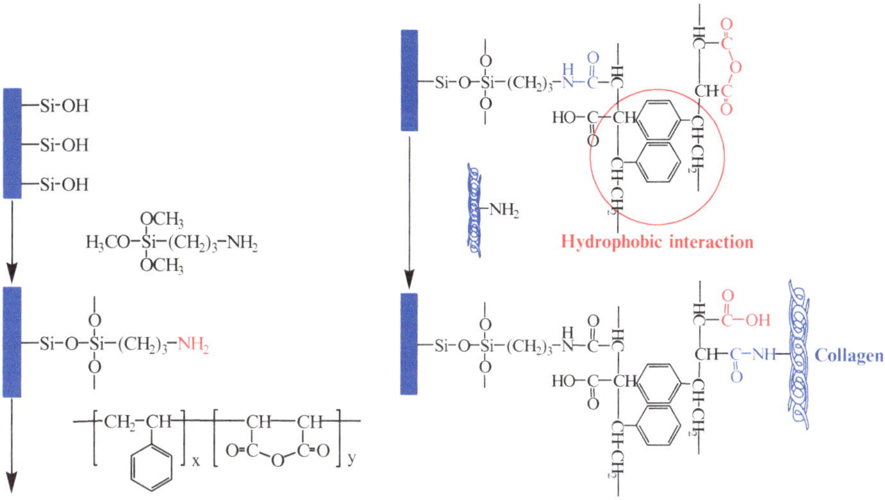

Fig. 3.12 Immobilization of ECM proteins by APTMS and PS-MA treatment

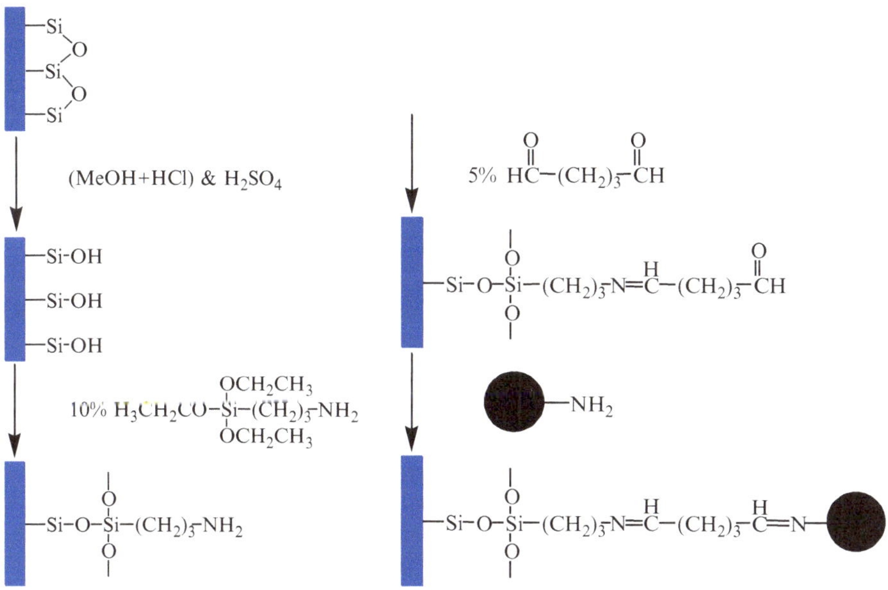

Fig. 3.13 Immobilization of ECM proteins by APTMS and glutaraldehyde treatment

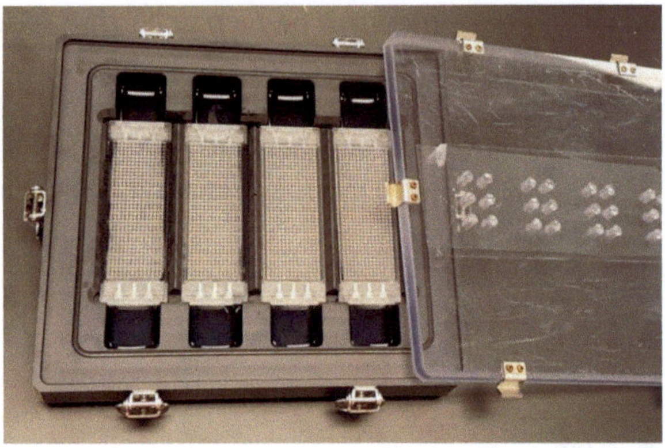

Fig. 3.14 Picture of a humid incubation chamber

1. Prepare ECM protein solutions (40 μg/200 μL) in sterile PBS (1×) and place them on ice. Typical ECM proteins used are as follows: Matrigel™, collagen I, fibronectin, laminin, vitronectin, *etc.*
2. Print 60 nL of ECM protein solutions on the reactive surfaces of glass slides using S+ MicroArrayer.
3. Incubate the slides for 1 h at room temperature in a humid incubation chamber to prevent evaporation of water (Fig. 3.14).
4. After drying the slides at room temperature in a biosafety cabinet, keep them in a sterile petri dish for long-term storage. It may be stored at 2–8 °C for up to several weeks under sterile conditions.
5. Before adding cell suspensions (e.g., Hep3B, MCF7, and A293 cells), rinse the slides in sterile PBS twice with shaking for 30 min at 100 rpm to remove unbound ECM proteins.
6. Culture the cells in proper growth media for 2–3 days in a CO_2 incubator at 37 °C to prepare cell patterns or confluent cell monolayers.

3.4 Summary

Glass and polystyrene are both suitable substrates for cellular microarrays, but both substrates require chemical modifications to ensure proper adhesion of cells or hydrogels. In the case of plastic chips made of polystyrene, it involves coating with amphiphilic, water-insoluble polymers. For glass slides, this means chemical modification with silanes, or spin-coating with amphiphilic polymers. In both cases, these only serve as preliminary steps to ensure adequate biocompatibility and cell/hydrogel attachment, as subsequent steps are usually necessary to generate microarrays. Nevertheless, these steps are necessary to create surfaces that can be used for biological assay development with living cells.

References

1. Lee, M.-Y., Kumar, R. A., Sukumaran, S. M., Hogg, M. G., Clark, D. S., & Dordick, J. S. (2008). Three-dimensional cellular microarray for high-throughput toxicology assays. *Proceedings of the National Academy of Sciences, 105*(1), 59–63. doi:10.1073/pnas.0708756105.
2. Ramsey, W., Hertl, W., Nowlan, E., & Binkowski, N. (1984). Surface treatments and cell attachment. *In Vitro, 20*(10), 802–808.
3. Curtis, A. S. G., Forrester, J. V., McInnes, C., & Lawrie, F. (1983). Adhesion of cells to polystyrene surfaces. *The Journal of Cell Biology, 97*(5I), 1500–1506. doi:10.1083/jcb.97.5.1500.
4. Amstein, C. F., & Hartman, P. A. (1975). Adaptation of plastic surfaces for tissue-culture by glow-discharge. *Journal of Clinical Microbiology, 2*(1), 46–54.
5. Hudis, M. (1974). Plasma treatment of solid materials. In R. H. Hansen & A. T. Bell (Eds.), *Techniques and applications of plasma chemistry* (p. 147). New York: John Wiley and Sons.
6. Seo, J. H., Shin, D. S., Mukundan, P., & Revzin, A. (2012). Attachment of hydrogel microstructures and proteins to glass via thiol-terminated silanes. *Colloids and Surfaces B: Biointerfaces, 98*, 1–6. doi:10.1016/j.colsurfb.2012.03.025.
7. Carvalho, R. R., Pujari, S. P., Lange, S. C., Sen, R., Vrouwe, E. X., & Zuilhof, H. (2016). Local light-induced modification of the inside of microfluidic glass chips. *Langmuir, 32*(10), 2389–2398. doi:10.1021/acs.langmuir.5b04621.
8. Luo, N., Zhong, H., Yang, M., Yuan, X., & Fan, Y. (2016). Modifying glass fiber surface with grafting acrylamide by UV-grafting copolymerization for preparation of glass fiber reinforced PVDF composite membrane. *Journal of Environmental Sciences, 39*, 208–217. doi:10.1016/j.jes.2015.11.010.
9. Cheawchan S, Uchida S, Sogawa H, Koyama Y, & Takata T. Thermotriggered catalyst-free modification of a glass surface with an orthogonal agent possessing nitrile *N*-Oxide and masked Ketene functions. *Langmuir* 2015;32(1): 309–315. doi:10.1021/acs.langmuir.5b03881. (Fig. 1).
10. Tang, J., Peng, R., & Ding, J. (2010). The regulation of stem cell differentiation by cell-cell contact on micropatterned material surfaces. *Biomaterials, 31*(9), 2470–2476. doi:10.1016/j.biomaterials.2009.12.006.
11. *User's manual for PDC-001-HP (115V) or PDC-002-HP (230V) high power expanded plasma cleaner (and optional PDC-FMG (115V) or PDC-FMG-2 (230V) PlasmaFlo)*. Ithaca, NY; 2015.
12. Cras, J. J., Rowe-Taitt, C. A., Nivens, D. A., & Ligler, F. S. (1999). Comparison of chemical cleaning methods of glass in preparation for silanization. *Biosensors & Bioelectronics, 14*(8–9), 683–688. doi:10.1016/S0956-5663(99)00043-3.
13. No Title. Retrieved from http://www-schreiber.chem.harvard.edu/home/protocols/SMP.htm.

References

Chapter 4
Biological Sample Printing

Parnian Bigdelou, Alexander Roth, Akshata Datar, and Moo-Yeal Lee

Contents

4.1 Introduction 71
4.2 Materials 73
4.3 Preparation of Stock Solutions 74
4.4 Daily Operation of the S+ MicroArrayer for Dispensing Biological Samples 76
4.5 Sample Printing Protocols 80
4.5.1 Enzyme Printing for Metabolism-Induced Drug Toxicity Assays 80
4.5.2 Compound Printing 82
4.5.3 Cell Printing 85
4.5.3.1 Preparation of Cell Suspension in Growth Medium 85
4.5.3.2 Cell Printing for 2D Monolayer Culture on the Micropillar Chip 86
4.5.3.3 Cell Printing in Alginate for 3D Cultures 87
4.5.3.4 Cell Printing in Matrigel for 3D Cultures 90
4.5.3.5 Cell Printing in a Mixture of Alginate and Matrigel for 3D Cultures 93
4.5.3.6 Cell Printing in a Mixture of Alginate and Fibrinogen for 3D Cultures 95
4.5.3.7 Cell Printing in PuraMatrix for 3D Cultures 96
4.5.4 Virus Printing 98
4.5.4.1 Measurement of Viral Titer in a 96-Well Plate 98
4.5.4.2 Adenoviral Transduction on the Micropillar/Microwell Chip 99
4.6 Inspection of Cells Printed on the Micropillar Chips Using a Bright-Field Microscope 101
4.7 Coefficient of Variation (CV) and Z' Factor for Assay Validation 102
4.8 Summary 103
References 103

4.1 Introduction

Microarray spotters equipped with solenoid valves are capable of printing a wide range of biological samples, including reagents, growth media, compounds, hydrogels, genes, proteins, viruses, and cells, for biochemical and cell-based assays. Solenoid valve-driven bioprinting has clear advantages over other printing technologies

M.-Y. Lee, *Microarray Bioprinting Technology*,
DOI 10.1007/978-3-319-46805-1_4

in terms of controlling sample volume dispensed and flexibility in the type of samples printed. Unlike robotic liquid dispensers that are more robust due to relatively large orifice sizes and dispensing volumes, microarray spotters must take precautions to prevent the clogging of solenoid valves and ceramic tips due to the extremely small sample volume printed. Therefore, it is essential to avoid dust and large precipitates (e.g., compound precipitates and filamentous microbes), test the viscosity of the material printed, and understand the mechanism of gelation for hydrogels used for cell encapsulation. In general, temperature-sensitive hydrogels such as Matrigel have a high risk of forming a gel spontaneously inside tubes, ceramic tips, and solenoid valves with a slight change in temperature. Therefore, a lower temperature must be maintained constantly across the chip-loading deck and the dispensing head. The clogging may result in replacement of tubes, solenoid valves, and ceramic tips, which can be an expensive affair. Gelation of hydrogels used for 3D cell culture has to be occurred in two steps to avoid clogging. Typically, crosslinkers/initiators are printed first on the micropillar/microwell chip platform and then hydrogels containing cells are dispensed on the top to form a gel. Therefore, ionic crosslinking (e.g., alginate with $CaCl_2$, PuraMatrix with salts), affinity/covalent bonding (e.g., functionalized polymers with streptavidin and biotin), photopolymerization (e.g., methacrylated alginate with photoinitiators), and biocatalysis (e.g., fibrinogen with thrombin) are favorable mechanisms of gelation. Developing proper surface chemistry to attach cell spots in hydrogels robustly on the surface of the micropillar/microwell chip is also critical. Covalent bonding (e.g., poly(maleic anhydride-*alt*-1-octadecene) and poly-L-lysine), affinity (e.g., streptavidin and biotin), and ionic interaction (poly-L-lysine and negatively charged alginate) are commonly introduced to enhance spot attachment on the chip. Viscosity of the material printed is another important factor that affects the performance of printing, and valve/tip clogging and rinsing. In general, biological samples with low viscosity such as 10–30 centipoise (equivalent to 1 % or lower alginate in distilled water) are preferred. Samples with high viscosity cannot be printed with solenoid valves and are difficult to remove from tubes, solenoid valves, and tips by rinsing with water. Therefore, it has to be diluted properly with either solvents or cell culture media to ensure reproducible printing. In case of cell printing and encapsulation for 3D culture, maintaining cell suspension in hydrogels while printing is important to avoid solenoid valves and tips clogging. In general, over crowded cells (typically seeding density higher than 10 million cells/mL) can result in clogging issues. In addition, cells tend to settle down quicker in a low viscosity solution. Finally, mechanical strength of hydrogels over time needs to be considered to support long-term cell culture. Peptide-based hydrogels (e.g., fibrinogen, Matrigel, and PuraMatrix) tend to lose their strength over time due to degradation by matrix metalloproteinases (MMPs) secreted by many mammalian cells. These MMPs are responsible for hydrogel degradation and eventual spot detachment. To sustain cell spots for a longer time and minimize spot detachment due to degradation, peptide-based hydrogels can be mixed with more stable hydrogels such as alginate. When mixed, the resulting hydrogel has to be transparent with minimal background fluorescence and should not interfere high-content imaging (HCI) assays. The protocols provided in this chapter will give researchers a guidance towards multiplex, 3D-cell based assays on the micropillar/microwell chip platform.

4.2 Materials

- Poly maleic anhydride-*alt*-1-octadecene (PMA-OD) (Sigma-Aldrich)
- Alginate (Alginic acid sodium salt; Sigma-Aldrich)
- Barium chloride, dehydrate ($BaCl_2$; Sigma-Aldrich)
- Poly-L-lysine, 0.01 % (PLL; Sigma-Aldrich)
- PureCol® bovine collagen solution, Type I, 3 mg/mL (Advanced Biomatrix)
- Matrigel (Corning)
- Fibrinogen from bovine plasma (Sigma-Aldrich)
- Thrombin from bovine plasma (Sigma-Aldrich)
- Dulbecco's phosphate-buffered saline (DPBS) without Ca^{2+} and Mg^{2+}
- 0.05 % Trypsin-EDTA (1×), phenol red (Life Technologies)
- 0.25 % Trypsin-EDTA (1×), phenol red (Life Technologies)
- Penicillin-Streptomycin, liquid (Life Technologies)
- Control Baculosomes® plus reagents (Life Technologies)
- P450-3A4 Baculosomes® plus reagents, rHuman (Life Technologies)
- P450-2D6 Baculosomes® plus reagents, rHuman (Life Technologies)
- P450-2C9 Baculosomes® plus reagents, rHuman (Life Technologies)
- P450-2E1 Baculosomes® plus reagents, rHuman (Life Technologies)
- P450-1A2 Baculosomes® plus reagents, rHuman (Life Technologies)
- P450-3A5 Baculosomes® plus reagents, rHuman (Life Technologies)
- P450-2C8 Baculosomes® plus reagents, rHuman (Life Technologies)
- P450-2C19 Baculosomes® plus reagents, rHuman (Life Technologies)
- P450-2B6 Baculosomes® plus reagents, rHuman (Life Technologies)
- Vivid® regeneration system, 100X (Life Technologies)
- UGT1A1, human, 5 mg/mL (Corning)
- UGT1A3, human, 5 mg/mL (Corning)
- UGT1A4, human, 5 mg/mL (Corning)
- UGT1A9, human, 5 mg/mL (Corning)
- UGT2B4, human, 5 mg/mL (Corning)
- UGT2B7, human, 5 mg/mL (Corning)
- UGT reaction mix A (containing 25 mM UDP-GA) (Corning)
- UGT reaction mix B (containing alamethicin) (Corning)
- Sulfotransferase isozyme 1A1 human (SULT1A1, Cypex)
- Sulfotransferase isozyme 1A3 human (SULT1A3, Cypex)
- Sulfotransferase isozyme 1B1 human (SULT1B1, Cypex)
- Glutathione *S*-transferase (GST) A1 human (Cypex)
- Glutathione *S*-transferase (GST) M1 human (Cypex)

- β-Nicotinamide adenine dinucleotide phosphate sodium salt (NADP; Sigma-Aldrich)
- Adenosine 3′-phosphate 5′-phosphosulfate lithium salt hydrate, >60 % (PAPS; Sigma-Aldrich)
- Glutathione reduced (GSH; Sigma-Aldrich)
- Micropillar/microwell chips (25 × 75 × 2 mm from SEMCO, Suwon, South Korea)
- Staining plates for micropillar/microwell chips (SEMCO, Suwon, South Korea)
- Round bottom 96-well plate (Corning)
- S+ MicroArrayer with six solenoid valves and ceramic tips (SEMCO, Suwon, South Korea)
- S+ Scanner with four filter sets (SEMCO, Suwon, South Korea)

Materials for Hepatocytes

- Human cryopreserved hepatocytes (ThermoFisher Scientific) ***Note: Cryopreserved hepatocytes must be stored in the vapor phase of a liquid nitrogen freezer at less than −150 °C to assure the long term viability of these cells. When stored under the recommended conditions, the cryopreserved hepatocytes may be stable for up to 5 years.***
- CHRM™ cryopreserved hepatocyte recovery medium (ThermoFisher Scientific)
- Thawing/plating supplements (ThermoFisher Scientific)
- Cell maintenance supplements (ThermoFisher Scientific)
- 1× DMEM with 4.5 g/L glucose and sodium pyruvate without L-glutamine and phenol red (Corning)
- Corning® 500 mL vacuum filter (Corning)
- Trypan blue solution, 0.4 % liquid, sterile-filtered, cell culture tested (Sigma-Aldrich)
- Matrigel (Corning). ***Note: Store Matrigel solution in aliquots at −20 °C and keep frozen until used. Do not store in a frost-free freezer. Avoid multiple freeze-thaws.***
- Dulbecco's phosphate-buffered saline (DPBS) without $CaCl_2$ & $MgCl_2$, 500 mL (ThermoFisher Scientific)

4.3 Preparation of Stock Solutions

Preparation of $CaCl_2$ Stock Solution

1. Prepare 200 mM $CaCl_2$ stock solution in sterile distilled water by dissolving 1.5 g of $CaCl_2 \cdot H_2O$ powder in 50 mL of deionized distilled water and passing it through a 0.2 μm syringe filter.
2. Store the stock solution in a 4 °C refrigerator until use.

Preparation of Matrigel® Stock Solution

1. Thaw Matrigel from Corning by storing the bottle in a 4 °C refrigerator overnight.
2. While cold, quickly vortex the bottle to ensure that Matrigel is evenly dispersed.
3. Place Matrigel on ice and take aliquots of 500 μL Matrigel with pre-cooled pipette tip. ***Note: Do not change the pipette tip as Matrigel is viscous and sticking to the side of the tip. It is important to avoid multiple freezing and thawing of the Matrigel stock solution.***
4. Immediately after taking aliquots in Eppendorf tubes, store them in a −20 °C freezer until use. ***Note: Aliquots of Matrigel should be prepared quickly and stored to prevent gelation.*** Gelled Matrigel may be re-liquefied if placed at 4 °C on ice for 24–48 h.

Preparation of Thrombin Stock Solution

1. Add 20.7 mg of thrombin in 20.3 mL of DPBS to prepare a stock solution of 100 NIH units/mL. Thrombin comes in 20.7 mg of lyophilized powder which is equivalent to 98 NIH units/mg of protein (i.e., 2028.6 NIH units of thrombin in total). ***Note: Thrombin can be dissolved in distilled water, but a buffer solution is preferred.***
2. Prepare aliquots of 500 μL of thrombin in DPBS to avoid multiple freezing and thawing of the enzyme stock solution.
3. Store the thrombin stock solution at −20 °C. ***Note: Aliquots of thrombin should be prepared quickly and stored to prevent autolytic digestion.***

Preparation of Fibrinogen Stock Solution

1. Prepare 0.9 % NaCl solution in distilled water and sterile filter it.
2. Warm the 0.9 % NaCl solution to 37 °C in a water bath.
3. Add 100 mg of fibrinogen in 1 mL of 0.9 % NaCl to prepare a stock concentration of 100 mg/mL.
4. Agitate the solution slightly by shaking to dissolve. If necessary, further dilute the stock concentration to obtain a clear solution. ***Note: Fibrinogen solution should not be vortexed. Fibrinogen solution cannot be stored, and should be prepared fresh every time.***

Preparation of $BaCl_2$ Stock Solution

1. Prepare a 0.1 M barium chloride ($BaCl_2$) solution by adding barium chloride dehydrate in sterile de-ionized distilled water and vortex well for complete dissolution.
2. After sterile filtering the solution using a 0.2 μm syringe filter, store the 0.1 M $BaCl_2$ solution in aliquots in a 4 °C refrigerator until use.

Preparation of Alginate Stock Solution

1. Prepare a 3 % (w/v) alginic acid solution by dissolving alginic acid sodium salt in sterile de-ionized distilled water in a sterile glass sample vial.
2. Place a small stir bar in the vial and let it mix for at least 3 days for complete dissolution.
3. Store the 3 % alginate solution in aliquots in a 4 °C refrigerator until use.

4.4 Daily Operation of the S+ MicroArrayer for Dispensing Biological Samples

1. Open the air cylinder (or in-house air valve) and maintain the pressure of compressed air at 100 psi.
2. Turn on the chiller, the external switch in the utility body, the computer, and the monitor.
3. Prior to running the 'ezAOI', push 'Reset button' and reset XYZ coordinates. ***Note: This step is essential to avoid malfunctioning of the microarray spotter. Do not skip this step!***
4. Run the 'ezAOI' software (Fig. 4.1). ***Note: Make sure that no obstacles are on the work place to avoid the robotic arm crashing.***
5. Go to 'Water Alcohol Change' in the 'User Operation' box (Fig. 4.2).
6. Click 'Release' in the 'Air pressure' box to release air pressure in the pressure bottles.

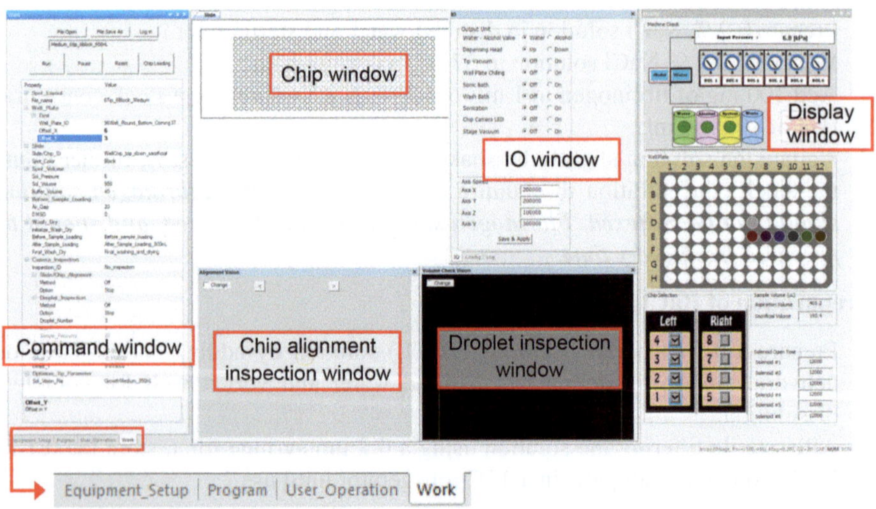

Fig. 4.1 The main screen of ezAOI. The Command window has four groups—Equipment setup, Program, User operation, and Work

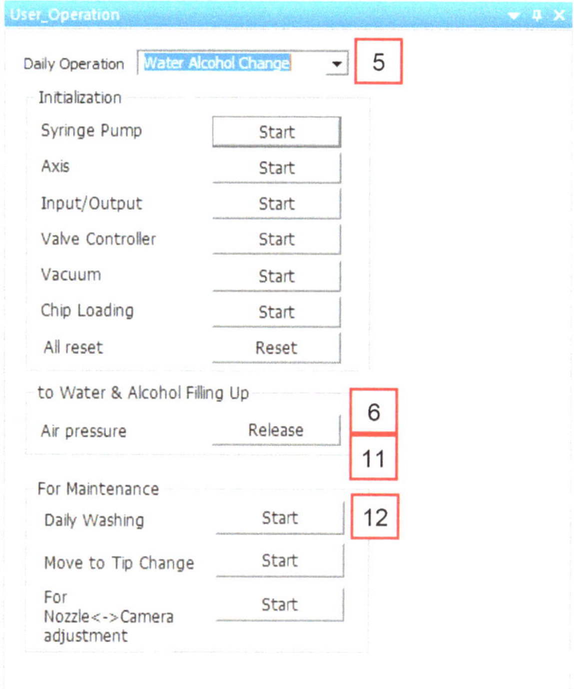

Fig. 4.2 Screen of 'User Operation' window > 'Water Alcohol Change'

7. Turn the orange knob connected to the pressure bottles from 'S' to 'O' position, which releases the pressure inside.
8. Open the lid of the pressure bottles and fill the pressure bottles with distilled water and 70 % ethanol.
9. Close the lid of the pressure bottles carefully. ***Note: Make sure that the 'O' ring is properly placed before placing the lid.***
10. Turn the orange knob of the pressure bottles from 'O' to 'S' position, which close the pressure bottles completely.
11. Click 'Initialize' of the 'Air pressure' box in the 'Water Alcohol Change' box to apply air pressure in the bottles.
12. Click the 'Start' button in the 'Daily Washing' box. ***Note: Washing tubes and solenoid valves with alcohol is essential to remove air bubbles prior to sample dispensing. After alcohol washing, thorough rinsing of tubes and solenoid valves with water is necessary to avoid enzyme deactivation or cell death due to remaining alcohol. An extensive rinse of tubes and solenoid valves with ice-cold water is critical before and after printing Matrigel solutions.***
13. Check a streamline of water while washing. ***Note: If the streamline is deflected from the vertical or water is beaded up at the tip end, clean the ceramic tip with sonication.***
14. Select the 'Work' window (Fig. 4.3).

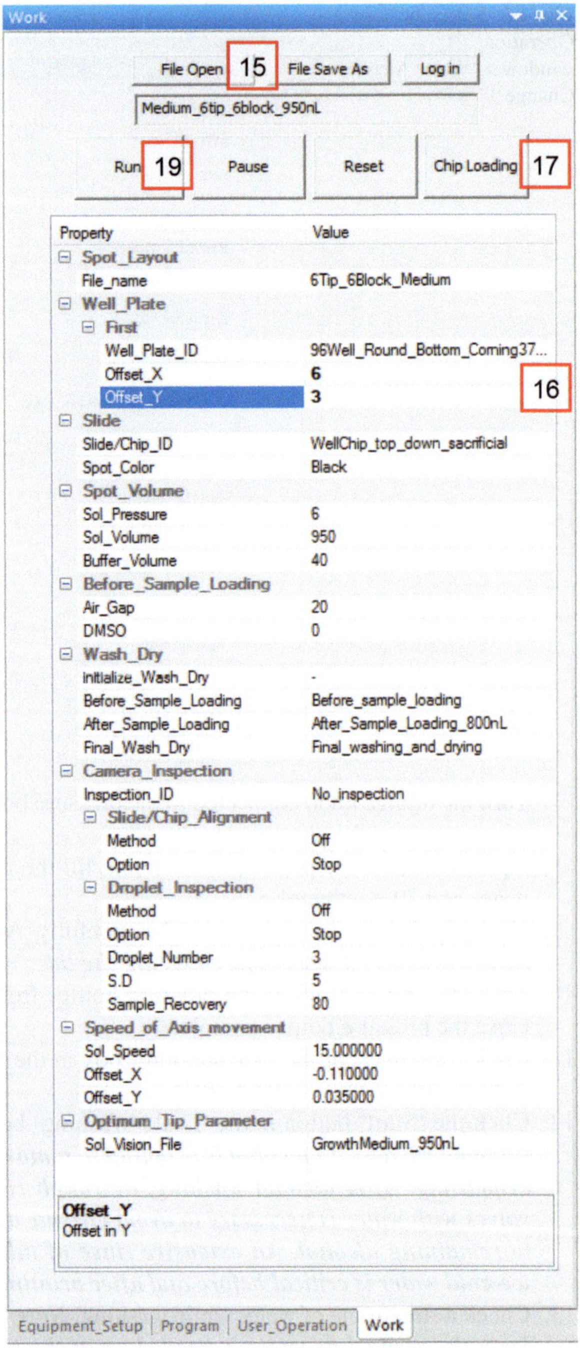

Fig. 4.3 Screen of 'Work' window

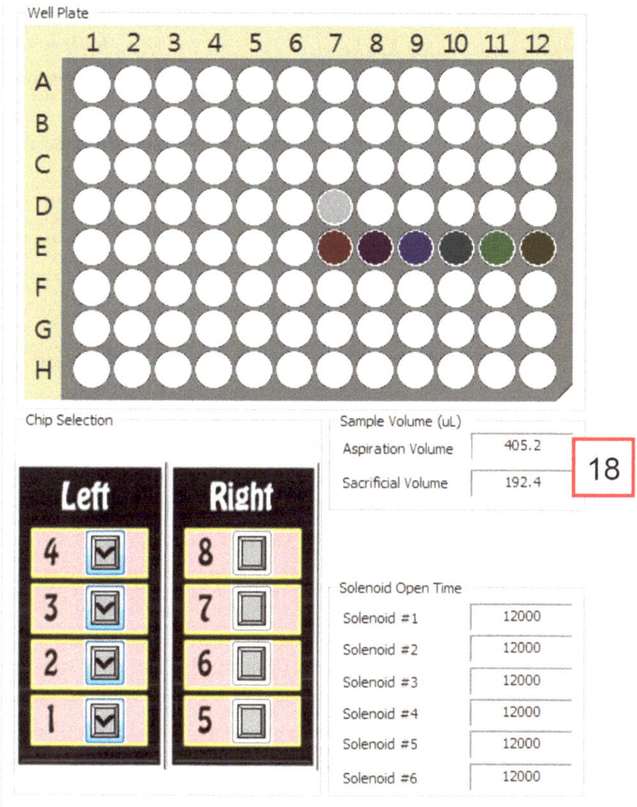

Fig. 4.4 Screen of 'Display' window

15. Click the 'File Open' button, and select the desired work file.
16. Enter the offset values in the X and Y directions to designate the location of samples in 96-wells. The exact location of sample wells will be appeared in the schematic of the well plate in the 'Display' window (Fig. 4.4).
17. Click the 'Chip Loading' button, and place the chips on the chilled chip deck shortly before sample printing and select the location of chips in the 'Display' window onto which samples are dispensed (Fig. 4.4).
18. Add proper amounts of samples based on aspiration and sacrificial volumes shown in the 'Display' window, in the designated wells, and place the well plate on the deck. ***Note: Always add 20 μL more to the aspiration and sacrificial volumes in the 96-well plate in case of evaporation.***
19. Click the 'Run' button.
20. To avoid potential contamination issues after sample printing, rinse tubes and solenoid valves with ethanol and water thoroughly by running 'Daily Washing'.
21. Close 'ezAOI' software.

22. Turn off the computer, the monitor, the external power switch in the utility body, and the chillers. ***Note: Make sure to turn off the chillers to avoid excess water condensation on the chip-loading deck and the dispensing head, causing a short circuit by water.***
23. Close the air cylinder (or in-house air valve). ***Note: Do not close the air cylinder while 'ezAOI' is on.***

4.5 Sample Printing Protocols

4.5.1 Enzyme Printing for Metabolism-Induced Drug Toxicity Assays

A miniaturized enzyme array (Metabolizing Enzyme Toxicology Assay Chip or **MetaChip**) has been prepared on the microwell chip to emulate the metabolizing reactions in the human liver [1–3]. The MetaChip is an array of drug metabolizing enzymes encapsulated in hydrogels, which is complementary to the micropillar chip containing hepatic cell lines cultured in 3D (Data Analysis Toxicology Assay Chip or **DataChip**) [1–3]. The MetaChip sandwiched with the DataChip has been used to identify metabolism-induced toxicity of compounds. Briefly, six different compounds can be printed in sections 1–6 of the MetaChip, each region containing a 3 × 6 mini-array (Fig. 4.5). Within each mini-array, six different doses of a compound can be assayed for toxicity. For on-chip drug metabolism, human metabolizing enzymes can be transversely printed into four regions (A–D), each containing no enzyme and different mixtures of enzymes encapsulated in Matrigel, thereby creating 24 distinct regions on the MetaChip (Table 4.1). Other human and rat metabolizing enzymes can be spotted similarly on the microwell chip for drug metabolism.

1. Thaw a Matrigel aliquot overnight in a refrigerator at 4 °C. ***Note: Matrigel may gel at slightly elevated temperatures in a refrigerator.***
2. Keep metabolizing enzymes and Matrigel on ice before use, and use pre-cooled pipettes, tips, and tubes when preparing a mixture of Matrigel and enzyme for spotting.
3. Turn on the chiller connected to the dispensing head, the chip-loading deck and the 96-well plate deck, set the temperature at 4–12 °C, and circulate cold water. ***Note: The chilling temperature should be low enough to avoid water evaporation in the spots. The chilling temperature may vary depending on the temperature and humidity of the room. If there is severe water condensation observed on the dispensing head or the chip-loading deck, then the chilling temperature is too low. Make sure to turn the chiller connected to the dispensing head off after Matrigel spotting to avoid extensive water condensation on the dispensing head, potentially causing a short circuit.***
4. Operate the daily washing program with ice-cold water and make sure that all tubing, solenoid valves, and ceramic tips are cold prior to Matrigel spotting.

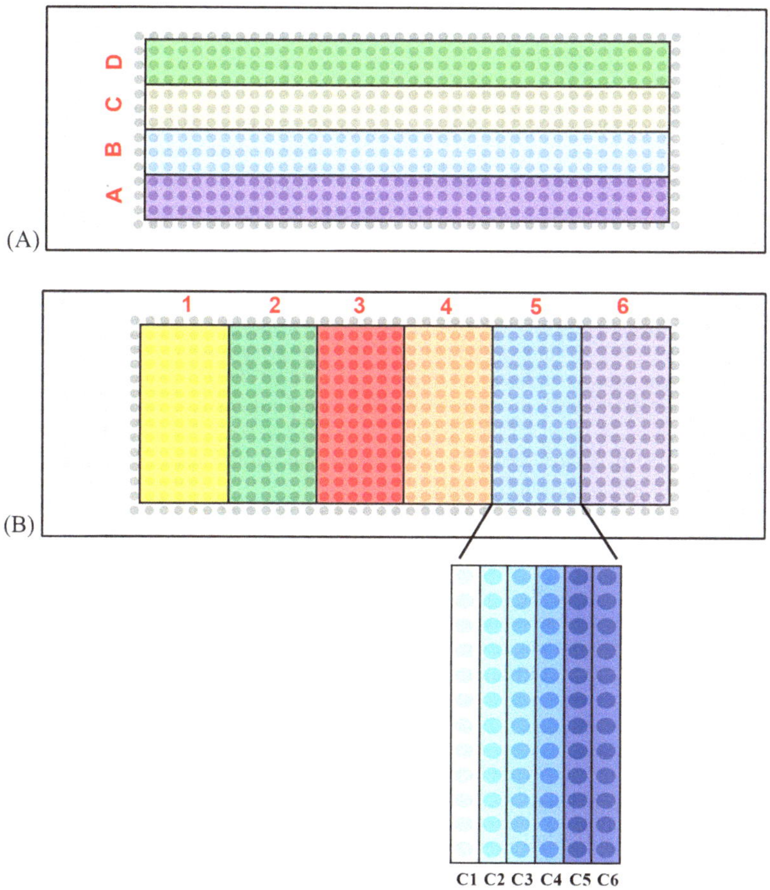

Fig. 4.5 (**a**) A layout of the MetaChip (432 spots/chip) for *in-situ* drug metabolism. Specifically, region A contains no enzyme as a test compound only control, regain B contains a mixture of human Cytochrome P450 isoforms, region C contains a mixture of human phase II drug-metabolizing enzymes and P450 enzymes (i.e., all enzyme mixture), and region D contains human liver microsome. (**b**) A layout of compounds printed in the MetaChip (432 spots/chip). Each different compound is printed in sections 1–6 of the MetaChip, and the concentrations (C1–C6) vary within each mini-array. The grey dots represent the microwells

At least 2 cycles of running the wash program with cold water is necessary to cool it down.

5. Prepare cold metabolizing enzyme solutions in Matrigel in a 96-well plate on ice (see Table 4.1 for details). ***Note: Always use the same kind of the well plates to avoid tip crashing due to Z height difference in wells. Always prepare fresh enzyme-Matrigel mixtures on ice and use them immediately.***

Table 4.1 Typical preparation of enzyme solutions for the MetaChip

Region	Enzymes	Co-substrates	Matrix
A	120 μL Insect cell microsomes control[a]	120 μL RPMI	80 μL Matrigel
B	120 μL P450 mixture[b]	120 μL Phase I cofactor[e]	80 μL Matrigel
C	120 μL All Mix[c]	120 μL All cofactor[f]	80 μL Matrigel
D	120 μL Human liver microsome[d]	120 μL All cofactor	80 μL Matrigel

[a]Insect cell microsomes (P450 negative control, 5 mg/mL) is purchased from BD biosciences and used without dilution

[b]P450 mixture contains 52 % 3A4 (1 μM), 20 % 2D6 (1 μM), 8 % 2C9 (1 μM), 5 % 2E1 (1 μM), 4 % 1A2 (1 μM), 4 % 3A5 (1 μM), 3 % 2C8 (1 μM), 3 % 2C19 (1 μM), and 1 % 2B6 (1 μM), all from Life Technologies

[c]All Mix contains 50 % P450 mixture and 50 % Phase II enzyme mixture

Phase II mixture contains 12 % UGT1A1 (5 mg/mL), 12 % UGT1A3 (5 mg/mL), 12 % UGT1A4 (5 mg/mL), 12 % UGT1A9 (5 mg/mL), 12 % UGT2B4 (5 mg/mL), 12 % UGT2B7 (5 mg/mL) from BD biosciences, 4 % SULT1A1 (250 μg/50 μL), 4 % SULT1A3 (250 μg/50 μL), 4 % SULT1B1 (250 μg/50 μL) from Cypex, 6 % GST (5 mg/200 μL) from Sigma, 5 % NAT1 (2.5 mg/mL), and 5 % NAT2 (2.5 mg/mL) from BD biosciences

[d]Human liver microsome purchased from BD biosciences and used without dilution

[e]Phase I cofactor solution contains 50 % NADP (10 mM) and 50 % regeneration system from Life Technologies kits.

[f]All cofactor solution contains 50 % Phase I cofactor and 50 % Phase II cofactor

Phase II cofactor solution contains 50 % UDP-GA (2 mL BD solution A + 3 mL BD solution B = 10 mM in 50 mM Tris-HCl buffer, pH 7.5), 20 % GSH (100 mM in 20 mM PBS buffer, pH 8), 20 % PAPS (25 mM in 20 mM PBS buffer, pH 8), and 10 % acetyl CoA (25 mM in 10 mM PBS buffer, pH 8)

6. Print metabolizing enzyme solutions in Matrigel into microwells (typically 432 spots/chip, 100 nL per spot, 1.5 mm well-to-well distance) placed on the cold chip deck. Maintain humidity under 50 % with the cooling chip deck to retard evaporation of water in the spots.
7. Immediately after enzyme printing, place the microwell chip with metabolizing enzymes in a petri dish (4 MetaChips in 150 mm-diameter petri dish) and store in a - 80 °C freezer until use. ***Note: Do not dry the enzyme spots, causing enzyme deactivation.***
8. Operate the daily washing program at least twice with ice-cold water to make sure that no Matrigel remains in tubing, solenoid valve, and ceramic tip.

4.5.2 Compound Printing

Test compounds can be printed in the MetaChip with drug metabolizing enzymes for assessing metabolism-induced toxicity or in the microwell chip (with no enzymes) for measuring parent compound toxicity. Typical printing volume of compounds in growth media on the chip is 800–950 nL. It is critical to minimize the use of organic solvents in the compound solutions. Typically, less than 0.5 % of DMSO in the final compound solutions is allowed. Thus, at least 200-fold dilution of compounds in growth media is necessary. In addition, compounds have to be either completely dissolved in growth media or form stable colloidal suspension. Big precipitates of compounds in growth media will clog solenoid valves and ceramic tips, resulting in unpredictable printing outcomes.

1. Dissolve a powder form of a compound in DMSO to prepare a compound stock solution at 200-fold higher concentrations than the desired final concentration. A typical concentration of the compound in DMSO is 50–300 mM. ***Note: Nonvolatile and water-miscible DMSO is the most commonly used as a solvent unless the compound is insoluble in DMSO. Check the Sigma-Aldrich website for solubility data of compounds in organic solvents.***
2. Add 20–40 mg of the compound in a glass scintillation vial (6 mL) and dispense a predetermined volume of DMSO. ***Note: The volume of DMSO added can be varied depending on the molecular weight and the toxicity of the compound. Typically, 200–2000 μL of DMSO is added to prepare a compound stock solution.***
3. Sonicate the compound in DMSO to completely dissolve it. The solution can be warmed up slightly to facilitate the dissolution process. ***Note: Do not vortex the compound in DMSO as the undissolved compound will be attached on the side wall of the glass vial, making it difficult to dissolve. Prior to placing the vial with the compound in a warm water bath, check the Sigma-Aldrich website for temperature stability of the compound. Typical temperature used is 40–70 °C.***
4. If the compound is insoluble in DMSO, further dilute the compound stock by adding more DMSO, or try different organic solvents. Ethanol or methanol is the most commonly used as an alternative of DMSO.
5. Test solubility of the compound in growth media by adding 2 μL of the compound in DMSO to 398 μL of growth media in an Eppendorf tube (1 mL) and mixing well with sonication for 1–2 min. ***Note: Do not vortex it when dissolving in growth media. Evaluate solubility at 200-fold dilution as the maximum allowable DMSO content is 0.5 %. The compound in growth media has to be completely dissolved or form stable colloidal suspension.***
6. Leave the Eppendorf tube with the compound in growth media for at least 30 min to see if big precipitates are formed. ***Note: The formation of compound precipitates must be avoided to prevent solenoid valve and ceramic tip clogging. In case of big precipitates formed, further dilute the compound stock with DMSO.***
7. After testing solubility in growth media, take aliquots of the compound stock in DMSO (typically 40 μL aliquot) and store them in a −20 °C or −80 °C freezer until use. ***Note: Do not freeze and thaw the compound stock in DMSO multiple times. The aliquot of the compound stock is for a single use for generating reproducible data. Compounds dissolved in volatile organic solvents cannot be aliquoted and stored due to evaporation issues.***
8. Prepare a compound-in-DMSO plate by thawing the frozen compound stock at room temperature and adding 40 μL of the compound stocks in a round-bottom 96-well plate (e.g., the 96-wells A6, B6, C6, D6, E6, and F6 in Fig. 4.6A).

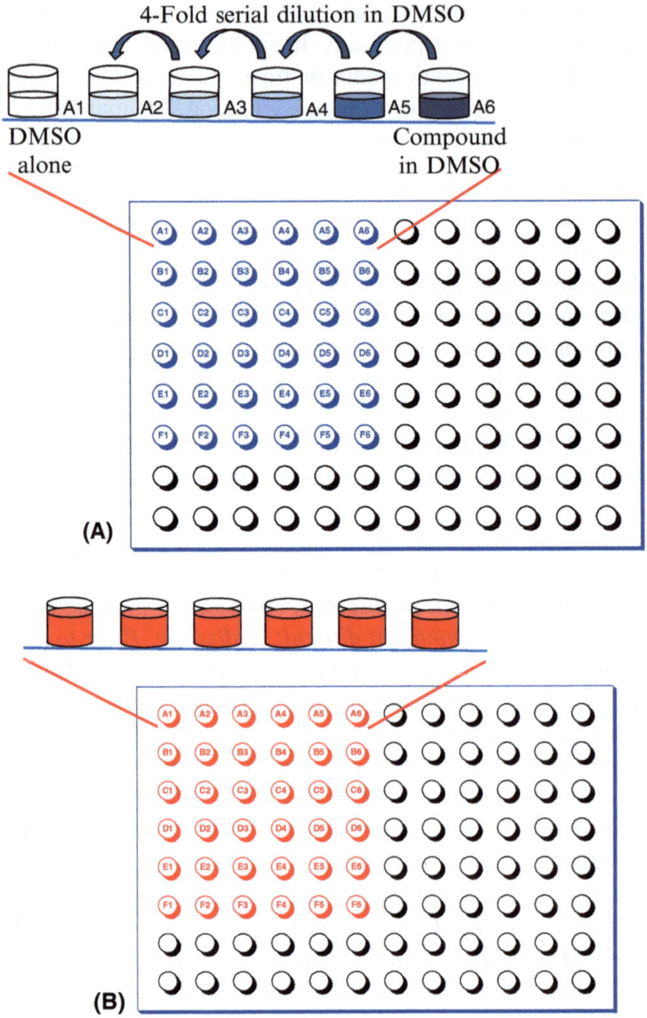

Fig. 4.6 (**a**) Preparation of a compound-in-DMSO plate in a 96-well plate by fourfold serial dilution. The 96-wells A6, B6, C6, D6, E6, and F6 contain 40 μL of compound stocks in DMSO and the remaining 96-wells containing 30 μL of DMSO. For fourfold serial dilution, 10 μL of compounds in DMSO are transferred to adjacent 96-wells and mixed well with a multi-channel pipette sequentially until the last 96-wells with DMSO alone remains. Thus, the 96-wells A1, B1, C1, D1, E1, and F1 containing DMSO alone are used as controls. (**b**) Preparation of a compound-in-growth medium plate in a 96-well plate. For 200-fold dilution of compounds in growth media, dispense 298.5 μL of growth media in the 96-wells A1–F6, transfer 1.5 μL of serially diluted compounds in the compound-in-DMSO plate to the 96-well plate containing growth media with a multi-channel pipette, and mix the solutions well by aspirating and dispensing at least 20 times. Samples in the 96-wells A1, B1, C1, D1, E1, and F1 are transferred first, and then sequentially from low to high concentrations to avoid compound carry-over

Note: Make sure of the complete dissolution of the compound in DMSO before serial dilution. In case of precipitates observed, sonicate or warm up the compound stock in DMSO.

9. For fourfold serial dilutions of compound stock solutions in the 96-well plate, add 30 μL of DMSO in other 96-wells, take 10 μL of the compound stock and transfer it to the adjacent 96-wells, and then mix the solution well by aspirating and dispensing the mixture with a multi-channel pipette at least 10 times (Fig. 4.6A).
10. This step is repeated sequentially until the last 96-wells with DMSO alone remains. Solvent alone (without compound) is used as a control (e.g., the 96-wells A1, B1, C1, D1, E1, and F1 in Fig. 4.6A). Typically, 6 compounds (each compound with 5 dosages and 1 solvent-alone control) are serially diluted from high concentrations to low concentrations in the 96-well plate.
11. Prepare a compound-in-growth medium plate by dispensing 298.5 μL of growth media in a round-bottom 96-well plate, adding 1.5 μL of serially diluted compounds in DMSO with a multi-channel pipette, and mixing the solution well by aspirating and dispensing (Fig. 4.6B). ***Note: To maintain final 0.5 % of DMSO in growth media, 200-fold compound dilution is used. Make sure to use the same kind of the 96-well plates to avoid tip crashing due to Z height difference in 96-wells.***
12. This step is repeated from low to high concentrations of compounds by transferring 1.5 μL of the serially diluted compounds in DMSO to the adjacent 96-wells containing 298.5 μL of growth media with a multi-channel pipette and then mixing the solution well by aspirating and dispensing at least 20 times. ***Note: Due to potential compound carry-over, always add the diluted compounds in DMSO from low to high concentrations in the 96-wells.***
13. Place the compound-in-growth medium plate and the microwell chips in S+ MicroArrayer.
14. Print 950 nL of compound solutions (typically 6 compounds, each with 5 dosages and 1 no compound control per chip) into the microwell chips using a work file (drug_6tip_36block) (Fig. 4.5B).

4.5.3 Cell Printing

4.5.3.1 Preparation of Cell Suspension in Growth Medium

It is very important to prepare well suspended cells in growth media prior to mixing with hydrogels for cell printing. Clumpy cells can clog solenoid valves and ceramic tips, resulting in unreliable printing problems. Always make sure to resuspend cells in the 96-well plate manually with a pipette before sample aspiration for uniform printing.

1. Carefully remove the medium from a tissue culture flask with approximately 80 % confluent monolayer of mammalian cells.
2. Rinse the flask once with DPBS. Add DPBS slowly from the side to avoid detaching the cells. Use 12 mL of DPBS for a T75 flask.
3. Aspirate out DPBS.
4. Add 1–2 mL of 0.05 % trypsin in the T75 flask, uniformly coat the cell monolayer with the trypsin solution, and incubate for 1–2 min at 37 °C.
5. Inspect the flask and ensure complete detachment of the cells by gently tapping the side of the flask with the palm of your hand.
6. Apply 5 mL of growth medium pre-warmed at 37 °C to the flask.
7. Aspirate and dispense the growth medium containing cells rigorously at least 20 times to break apart big cell clumps. ***Note: The trypsin step and this step are critical to prepare good cell suspension as clumpy cells can clog solenoid valves and ceramic tips.***
8. Gently rotate the flask to suspend the cells in growth medium.
9. Transfer the detached and suspended cells to a 15 mL conical tube.
10. Centrifuge the tube at 1200 rpm (typically 200 g) for 4 min to pellet the cells.
11. Aspirate out the entire supernatant gently without disturbing the cell pellet.
12. Add 2 mL of a complete growth medium supplemented with 10 % FBS and 1 % Pen/Strep and resuspend the cells thoroughly with a 1 mL pipette.
13. While cells are in good suspension, take 5 μL of the cell suspension out and mix it with 495 μL of growth media in an Eppendorf tube (1.5 mL) for cell counting at 100-fold dilution.
14. After well mixing with a pipette, take 75 μL of cell suspension out from the Eppendorf tube and load it in a Moxi Z cassette.
15. Insert the cassette in a Moxi Z mini automated cell counter to measure cell density in the number of cells per milliliter. Actual cell density is 100-fold higher than this density because of the dilution factor. ***Note: Do not vortex the cell suspension for mixing as mammalian cells are fragile.***
16. Dilute the cell suspension with growth media at a desired seeding density (typically, 6–10 million cells/mL) and then mix it with a hydrogel solution. The final cell seeding density is typically in the range of 1–4 million cells/mL, varied depending on printing volume on the chip and doubling time of the cells printed. The required volume of cell suspension can be calculated from the program (ezAOI) in S+ MicroArrayer by selecting an appropriate work file and the number of chips necessary.

4.5.3.2 Cell Printing for 2D Monolayer Culture on the Micropillar Chip

A variety of cell-based assays have been carried out on two-dimensional (2D) monolayers of mammalian cells because of straightforward method of preparation, low cost, and ease of use, compared to 3D cell culture [4]. Thus, 2D cell monolayer culture in 96-well plates has been widely used as a standard tool for high-throughput screening of drug candidates. Similar 2D cell culture has been demonstrated on the micropillar chip, and the protocol of cell printing is summarized below.

1. Coat the micropillar chip with 2 mL of 0.01 % (w/v) PMA-OD in ethanol, and let it dry at room temperature at least for 2–3 h (see the Sect. 3.3.2 in Chap. 3).
2. Prepare 0.0033 % poly-L-lysine (PLL) by mixing 0.01 % PLL stock (Sigma-Aldrich) with sterile deionized water at 1:2 ratio to obtain a final concentration of 0.0033 % PLL. ***Note: In order for robust cell attachment onto the micropillar chip, in addition to PLL, we may need collagen or laminin coating for some cells.***
3. Print 60 nL of 0.0033 % poly-L-lysine (PLL) on the micropillar chip using a work file and dry it for 2–3 h. ***Note: Coating the micropillar chip with biocompatible polymer solutions is critical for robust attachment of cells on the surface.***
4. Prepare cell suspension in the complete growth medium at a desired seeding density (1–6 million cells/mL).
5. Print 950 nL of the cell suspension in the microwell chips. ***Note: It is important to resuspend the cells in the growth medium immediately before printing. Printing cell suspension at too high seeding density may cause clogging of solenoid valves and ceramic tips.***
6. Sandwich the micropillar chip with polymer coating with the microwell chip containing cell suspension.
7. Place the sandwiched chips with the microwell chip on the top in a humidified petri dish with 10 mL of distilled water so that the cells can be settled on the surface of the micropillars.
8. To robustly attach the cells on the surface of the micropillars, incubate the sandwiched chips in the humidified petri dish in a 5 % CO_2 incubator at 37 °C for 6 h. ***Note: It is extremely important to prevent water evaporation during cell incubation as drying can result in cell death.***
9. After 6 h incubation, flip the sandwiched chips around to place the micropillar chip on the top, which may prevent leaching the growth medium out.
10. Incubate the sandwiched chips overnight for cell growth.
11. After robust cell attachment on the micropillar chip, change the growth medium every 1–2 days by printing fresh growth medium in the microwell chip, discarding the microwell chip containing old growth medium, and sandwiching the freshly prepared microwell chip. ***Note: The cells are cultured on the micropillar chip until 80–90 % confluency.***

4.5.3.3 Cell Printing in Alginate for 3D Cultures

Since 2D-cultured cells rapidly lose phenotypic properties and functions of tissues and may not accurately emulate the microenvironment and cellular architecture found *in vivo*, 3D cell culture strategies have been developed in efforts to foster retention of phenotypic tissue functions. We have successfully demonstrated human cell cultures in 3D on the micropillar/microwell chip platform using various hydrogels, including alginate, PuraMatrix™, fibrinogen, Matrigel, Geltrex, and photo-crosslinkable, methacrylated alginate. In particular, alginate is an ideal printable scaffold for tissue engineering because it is a naturally-derived hydrogel that is non-toxic, non-immunogenic, and non-adhesive in nature [5]. Alginate is also able to deliver soluble signaling molecules to cells, act as a support structure for cell growth and function, and provide space filling for future tissue ingrowth after hydrogel degradation [5]. Cells can easily be encapsulated and remain viable in alginate hydrogels, and it has been widely used in a variety of tissue engineering applications [5]. Alginate is a negatively charged biopolymer which forms a gel when interacting with a divalent cation such as Ca^{2+}, Ba^{2+}, Mg^{2+}, *etc.* Some divalent cations such as Ba^{2+} can be relatively toxic to cells, thus requiring additional rinsing of cells to remove excess Ba^{2+} ions. Alginate can be functionalized with methacrylate groups for photopolymerization [5, 6]. Photocrosslinkable alginate has been used as a biodegradable hydrogel system with independently adjustable physical, cell adhesiveness, and bioactive factor binding properties [7, 8]. Optimization of photoinitiators and their concentrations are necessary to minimize basal toxicity of cells printed. The protocol of cell printing in alginate using S+ MicroArrayer is summarized below.

1. Coat the micropillar chip with 2 mL of 0.01 % (w/v) PMA-OD in ethanol and dry it at room temperature at least for 2–3 h.
2. Prepare a mixture of 0.0033 % PLL and 25 mM $CaCl_2$ by mixing 0.01 % PLL with 37.5 mM $CaCl_2$ in distilled water at 1:2 ratio. ***Note: For human cell lines, a mixture of 0.0033 % PLL and 16 mM*** $BaCl_2$ ***can be used to improve spot attachment.*** $BaCl_2$ ***is much stronger crosslinker, but more toxic against some primary human cells and delicate cells.***
3. Print 60 nL of the PLL-$CaCl_2$ mixture on the micropillar chip using a work file (60 nL PLL_$CaCl_2$) and dry it for at least 2–3 h.
4. Print 950 nL of growth medium in the microwell chip placed on the chilling chip deck at 4–12 °C using a work file (medium_6tip_6block_950nL). ***Note: Turn on the chiller at least 1 h before sample printing.***
5. Immediately after printing, incubate the microwell chips with growth medium in a humid chamber with water to avoid water evaporation (Fig. 4.7).
6. Prepare 1–4 million cells/mL in 0.75 % alginate by mixing 1 mL of 2–8 million cells/mL, 0.5 mL of 3 % alginate in distilled water, and 0.5 mL of growth medium using a 1 mL pipette. ***Note: Do not vortex the mixture as mammalian cells are***

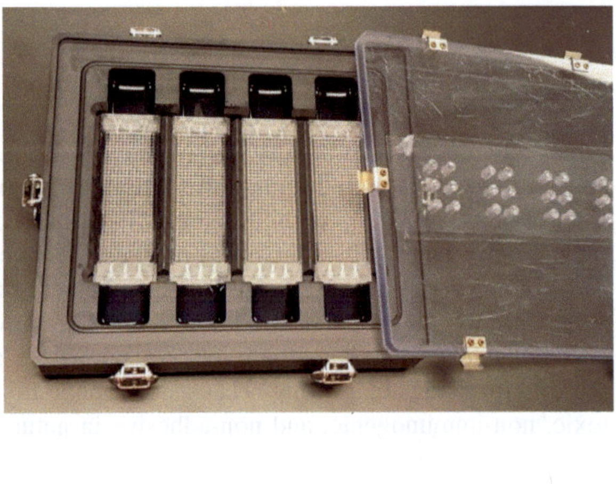

Fig. 4.7 Picture of a humid chamber with water. The lid will be put in place, and the chamber will be completely sealed to prevent water evaporation from growth medium in the microwell chip

fragile. Mix well by aspirating and dispensing the solution repeatedly (20 times) with a 1 mL pipette.

7. Place the micropillar chips with dried PLL-$CaCl_2$ spots on the chilling chip deck at 4–12 °C. ***Note: This is one of the most important steps in cell printing. Do not leave the micropillar chips too long on the chilling chip deck unattended as there might be severe water condensation observed on the chip surface, which will affect spot attachment. The chilling temperature should be adjusted to prevent spot drying as well as to avoid severe water evaporation on the surface.***
8. Add 100–200 μL/96-well of 1–4 million cells/mL in 0.75 % alginate in a 96-well plate. ***Note: Always use the same kind of the 96-well plates to avoid tip crashing due to Z height difference in 96-wells.***
9. Immediately before cell printing, resuspend the cells in 0.75 % alginate with a pipette. ***Note: This is one of the most important steps in cell printing. For uniform cell printing and preventing solenoid valve and tip clogging, it is extremely important to resuspend the cells immediately before printing using a 100 μL pipette. There might be severe clogging observed if cells are clumpy in alginate or the cell seeding density is too high (above 8 million cells/mL).***
10. Print 60 nL of the cell suspension in 0.75 % alginate on the micropillar chips placed on the chilling chip deck at 4–12 °C using a work file (Cell_Alginate_6 tip_6 block_60nL).
11. After 1–2 min gelation on the chilling chip deck, sandwich the micropillar chip with encapsulated cells with the microwell chip containing 950 nL of growth medium.
12. Place the sandwiched chips with the micropillar chip on the top in a humid cell incubation chamber or a petri dish with 10 mL water (Fig. 4.8B) for 30 min at 37 °C to remove excess $CaCl_2$. ***Note: In case of $BaCl_2$ used for alginate gelation, double rinsing with fresh growth medium might be necessary every 30 min.***
13. Discard the microwell chip and sandwich the micropillar chip with encapsulated cells with the microwell chip containing 950 nL of fresh growth medium.

(A)
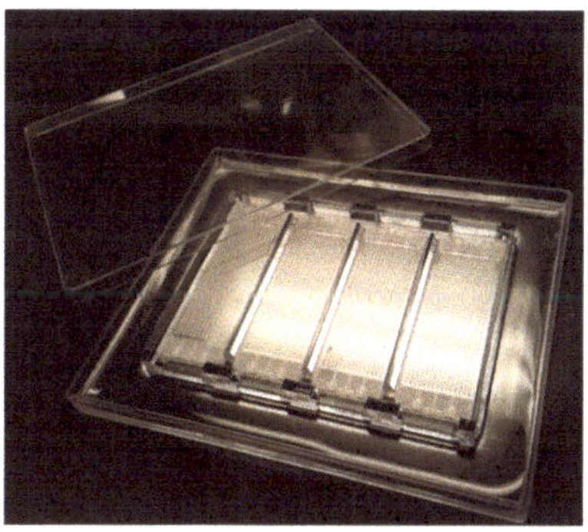
(B)
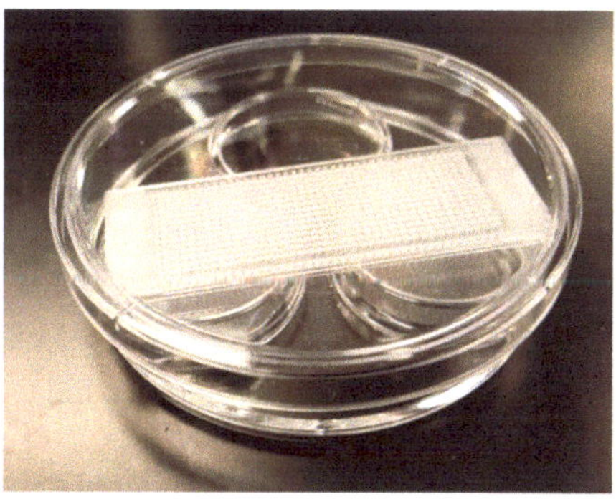

Fig. 4.8 Culture of cells on sandwiched chips in (**a**) a humid incubation chamber with 25 mL water and (**b**) a petri dish with 10 mL water

14. Inspect the sandwiched chips under a bright field microscope with 4× or 10× objective lens for spot attachment and uniform cell printing. ***Note: Make sure to avoid water evaporation during chip inspection. Do not leave out the sandwiched chips longer than 2 min for chip inspection. Water evaporation will change the concentration of salts in growth medium, thus influencing osmotic pressure inside cells and reducing cell viability and growth.***
15. Incubate the sandwiched chips in the humidified petri dish placed in a 5 % CO_2 incubator at 37 °C for 3D cell culture.
16. Change growth medium every 1–2 days by printing fresh growth medium in the microwell chip, discarding the microwell chip containing old growth medium, and sandwiching the freshly prepared microwell chip.

4.5.3.4 Cell Printing in Matrigel for 3D Cultures

Matrigel consists of laminin, collagen IV, entactin, heparin, and entactin which is produced by Engelbreth-Holm-Swarm (EHS) mouse sarcoma cells. It has been widely used in cell growth, differentiation, angiogenesis, and tissue vascularization, since it resembles cellular microenvironment in various tissues [9, 10]. Although Matrigel can be used for cell printing and encapsulation on the chip platform, extra caution has to be exercised as Matrigel is a temperature-sensitive hydrogel that forms a gel at a temperature ranging from 24 to 37 °C [11]. Therefore, it is necessary to maintain a low temperature on the chip-loading deck and in the dispensing head while printing Matrigel to avoid premature gelation and clogging in solenoid valves and ceramic tips. Typically, cells are mixed with cold Matrigel and printed immediately onto the micropillar chip. The printed cells in Matrigel can be gelled at 37 °C within 30 min [12]. Specific example of primary hepatocyte printing in Matrigel is presented below.

Preparation of Growth Media

1. Freshly prepare a thawing/plating medium by mixing 100 mL of 1× DMEM, 5 mL of FBS, 10 μL of dexamethasone, and 3.6 mL of cocktail A.
2. Freshly prepare a maintenance medium by mixing 100 mL of 1× DMEM, 1 μL of dexamethasone, and 4 mL of cocktail B.
3. Sterile filter the thawing/plating and maintenance media using the 0.2 μm filter kit from VWR.

Preparation of Primary Hepatocyte Suspension

1. Warm up the thawing/plating and CHRM™ media for at least 30 min in a 37 °C water bath before use.
2. Take out cryopreserved human hepatocytes in cryo-vial from a liquid nitrogen storage tank, release high nitrogen pressure in the cryo-vial by slightly unscrew-

ing the cap and then retightening it, and remove the paper label so that the ice crystal melting can be easily observed during the thawing process.

3. Immerse the cryo-vial containing hepatocytes in a 37 °C water bath and gently shake the cryo-vial in water until the last ice crystals floated to the top of the cryo-vial. ***Note: Do not take the vial out of water during the thawing process to check ice crystal melting. Always check it in water. The whole thawing process typically takes about 3 min.***
4. Immediately pour the melted hepatocytes from the cryo-vial into 50 mL of the CHRM™ medium.
5. Rinse the cryo-vial three times with 1 mL of the CHRM™ medium to recover hepatocytes attached on the wall.
6. Centrifuge the hepatocyte suspension in the CHRM™ medium at 100 g (RCF) and 20 °C for 10 min with the slowest brake and acceleration settings available on the centrifuge. ***Note: This is a critical step as primary hepatocytes are delicate.***
7. Aspirate out the supernatant into a waste bottle, leaving the hepatocyte pellet at the bottom.
8. Pour 3 mL of the thawing/plating medium onto the hepatocyte pellet and resuspend the pellet by gently swirling the tube or by tapping on the hood surface or by dragging the bottom of the tube over corrugated surface. ***Note: Do not use a vortex or a pipette to make well suspension of hepatocytes as they are very fragile. Do not shake the hepatocyte suspension.***
9. Split the thawing/plating medium containing suspended hepatocytes into two in Eppendorf tubes using a 1 mL pipette.
10. Centrifuge the hepatocyte suspension in thawing/plating medium again under the conditions stated above.
11. Aspirate out the supernatant into a waste bottle, leaving the hepatocyte pellet at the bottom.
12. Add 1.5 mL of the thawing/plating medium onto the hepatocyte pellet and resuspend the resulting hepatocyte pellet by gentle swirling. ***Note: Primary hepatocytes are extremely fragile. Do not vortex, shake, or aspirate/dispense hepatocytes to make the suspension.***
13. Combine two hepatocyte suspensions into one and centrifuge the combined hepatocyte suspension in thawing/plating medium under the conditions stated above.
14. Aspirate out the supernatant into a waste bottle, leaving the hepatocyte pellet at the bottom.
15. Add 300 μL of the thawing/plating medium onto the hepatocyte pellet, resuspend the resulting hepatocyte pellet by gentle swirling, and then store the hepatocyte suspension (8 million cells/mL) on ice until use.
16. To count the total number of hepatocytes, remove 10 μL of the hepatocyte suspension, add 490 μL of the thawing/plating medium, and measure the cell number in suspension using an automated cell counter.
17. To measure the viability of hepatocytes, remove 10 μL of the hepatocyte suspension, add 40 μL of the thawing/plating medium, mix the diluted hepatocyte suspension with 50 μL of trypan blue, and place 10 μL of the hepatocyte-trypan blue mixture on a hemocytometer.

Preparation of the Micropillar Chip with Primary Hepatocytes

1. Warm up the thawing/plating and maintenance media for at least 30 min in a 37 °C water bath before use.
2. Prior to use, thaw Matrigel overnight in a 4 °C refrigerator. Matrigel may gel at slightly elevated temperatures in the refrigerator.
3. Keep DPBS, Matrigel, and hepatocytes suspension on ice before use, and use pre-cooled pipettes, tips, and tubes when preparing a mixture of Matrigel and DPBS/cell suspension for spotting.
4. Turn on both chillers connected to the chilling chip deck, the dispensing head, the 96-well plate deck, and the water/alcohol containers at least 1 h before sample printing, set the temperature at 4 °C, and then circulate cold water to the dispensing head. ***Note: It is particularly important to maintain low temperature when printing Matrigel. Wait until the temperatures reach 4 °C. Make sure to turn off the chiller connected to the dispensing head after Matrigel spotting due to severe water condensation.***
5. Operate the daily washing program with ice-cold water and make sure that all tubes, solenoid valves, and ceramic tips are cold prior to Matrigel spotting. ***Note: At least running 2 cycles of the wash program with cold water are necessary to cool it down.***
6. Prepare a thawing/plating microwell chip by printing the thawing/plating medium into the microwell chip (typically 532 spots/chip, 950 nL per spot into microwell, 1.5 mm well-to-well distance).
7. Add 20 mL of sterile distilled water to the surrounding trench of the gas-permeable incubation chamber to retard water evaporation during incubation (Fig. 4.8A).
8. Place the thawing/plating microwell chip with 950 nL of the medium in the gas-permeable chamber and incubate it in a 5 % CO_2 incubator at 37 °C prior to stamping.
9. Prepare a Matrigel-DPBS mixture on ice by mixing 200 μL of cold Matrigel with 200 μL of cold Dulbecco's phosphate-buffered saline (DPBS).
10. Add approximately 500 μL of the cold Matrigel-DPBS mixture in a shallow-well plate (Fig. 4.9) and spread the solution uniformly on the surface of the shallow well using a pipette tip.

Fig. 4.9 Micropillar chip stamped on the shallow-well plate containing 500 μL of cold Matrigel-DPBS mixture

11. Sandwich the micropillar chip (typically 532 spots/chip, 1.5 mm pillar-to-pillar distance) with the shallow-well plate containing the cold Matrigel-DPBS mixture and then dry Matrigel-DPBS spots on the micropillar chip for 1 h in a sterile petri dish with a lid slightly open to yield flat Matrigel layers.
12. Prepare a suspension of hepatocytes in Matrigel on ice by mixing 300 μL of the hepatocyte suspension with 300 μL of Matrigel so that the final concentration of cells will be 4 × 10^6 cells/mL. ***Note: To prepare well-suspended hepatocytes in Matrigel, gently mix them on ice by aspirating and dispensing for 10 times. Don't mix cell suspension with Matrigel by vortex as some cell membranes are very fragile.***
13. Print the cold Matrigel solution containing hepatocytes (532 spots/chip, 50 nL per spot) atop each Matrigel-DPBS spot at 4 °C. ***Note: Put the 96-well plate with the cold Matrigel-hepatocyte mixture back on ice immediately after aspiration, or chill the 96-well plate deck by turning on the chiller and setting the temperature at 4 °C.***
14. Incubate the micropillar chip with hepatocytes in the black humid chamber with 700 μL water in each well at 37 °C for 20 min, resulting in gelation of the Matrigel matrix (Fig. 4.7). ***Note: Ensure the Matrigel gelling condition in your lab by adjusting the incubation time at 37 °C.***
15. If necessary, add freshly Matrigel solution containing suspended hepatocytes in the round-bottom 96-well plate to make sure well suspension of hepatocytes while spotting. After printing every chips (i.e., after every 5–10 min of spotting), add freshly Matrigel solution containing suspended hepatocytes in the well repeatedly. ***Note: Remind that hepatocytes are gradually settling down while spotting. For uniform cell spotting, keep re-suspending hepatocytes. After cell spotting, operate the daily washing program at least twice with ice-cold water to make sure that no Matrigel remains in tubes, solenoid valves, and ceramic tips.***
16. After incubation of the micropillar chip for Matrigel gelation, stamp the micropillar chip with hepatocytes onto the microwell chip with 950 nL of the thawing/plating medium, place the stamped chips in the gas-permeable chamber, and then incubate it in a 5 % CO_2 incubator at 37 °C for 2 h prior to compound toxicity tests.
17. Prepare compound solutions in the maintenance medium, print the medium in the microwell chip, stamp the micropillar chip with hepatocytes onto the microwell chip with compound solutions, and then incubate the stamped chips in a 5 % CO_2 incubator for 24 h for acute toxicity assays.
18. Stain hepatocytes on the micropillar chip with a Live/Dead® viability/cytotoxicity kit from Life Technologies.

4.5.3.5 Cell Printing in a Mixture of Alginate and Matrigel for 3D Cultures

Matrigel can be mixed with alginate to enhance cell viability and growth in alginate, while avoiding temperature-induced gelation. Thus, the concentration of Matrigel in alginate is substantially lower than its counterpart, Matrigel alone. The general protocol of cell printing in a mixture of alginate and Matrigel is presented below.

1. Coat the micropillar chip with 2 mL of 0.01 % (w/v) PMA-OD in ethanol and dry it at room temperature at least for 2–3 h.
2. Prepare a mixture of 0.0033 % PLL and 25 mM $CaCl_2$ by mixing 0.01 % PLL with 37.5 mM $CaCl_2$ in distilled water at 1:2 ratio.
3. Print 60 nL of the PLL-$CaCl_2$ mixture on the micropillar chip using a work file (60 nL PLL_$CaCl_2$) and dry it for at least 2–3 h.
4. Print 950 nL of growth medium in the microwell chip placed on the chilling chip deck at 4–12 °C using a work file (medium_6 tip_6 block). ***Note: It is particularly important to maintain low temperature when printing Matrigel. Turn on both chillers connected to the chilling chip deck, the dispensing head, the 96-well plate deck, and the water/alcohol containers at least 1 h before sample printing. Wait until the temperatures reach 4–12 °C. Matrigel is gelled at 37 °C.***
5. Immediately after printing, incubate the microwell chips with growth medium in a humid chamber with water to avoid water evaporation (Fig. 4.7).
6. Take out one aliquot of 500 μL Matrigel from a −20 °C freezer and thaw it overnight in a 4 °C refrigerator.
7. Place Matrigel on ice to avoid premature gelation and dilute the Matrigel stock solution (typically 9 mg/mL) with cold growth medium to obtain a Matrigel working solution (2 mg/mL).
8. Prepare 1–4 million cells/mL in 0.75 % alginate containing 0.5 mg/mL Matrigel by mixing 1 mL of 2–8 million cells/mL, 0.5 mL of 3 % alginate in distilled water, and 0.5 mL of Matrigel working solution (2 mg/mL) using a 1 mL pipette. ***Note: Do not vortex the mixture as mammalian cells are fragile. Mix well by aspirating and dispensing the solution repeatedly (20 times) with a 1 mL pipette.***
9. Place the micropillar chips with dried PLL-$CaCl_2$ spots on the chilling chip deck at 4–12 °C. ***Note: Do not leave the micropillar chips too long on the chilling chip deck unattended as there might be severe water condensation observed on the chip surface, which will affect spot attachment.***
10. Add 100–200 μL/96-well of 1–4 million cells/mL in 0.75 % alginate containing 0.5 mg/mL Matrigel in a 96-well plate. ***Note: Always use the same kind of the 96-well plates to avoid tip crashing due to Z height difference in 96-wells.***
11. Immediately before cell printing, resuspend the cells in 0.75 % alginate containing 0.5 mg/mL Matrigel with a pipette. ***Note: For uniform cell printing and preventing solenoid valve and tip clogging, it is extremely important to resuspend the cells immediately before printing using a 100 μL pipette.***
12. Print 60 nL of the cell suspension in 0.75 % alginate containing 0.5 mg/mL Matrigel on the micropillar chips placed on the chilling chip deck at 4–12 °C using a work file (Cell_Alginate_6 tip_6 block_60nL).
13. After 1–2 min gelation on the chilling chip deck, sandwich the micropillar chip with encapsulated cells with the microwell chip containing 950 nL of growth medium.
14. Rinse tubing, solenoid valves, and ceramic tips with cold distilled water by running the daily washing program twice. ***Note: Although Matrigel cannot be***

gelled in alginate at this low concentration, rinsing with cold water can ensure to remove residual Matrigel.

15. Place the sandwiched chips with the micropillar chip on the top in a humid cell incubation chamber or a petri dish with 10 mL water (Fig. 4.8B) and incubate the chips in a 5 % CO_2 incubator at 37 °C for 3D cell culture. ***Note: Unless cytotoxicity noticed due to $CaCl_2$, there will be no need to rinse the micropillar chip with growth medium.***
16. If necessary, inspect the sandwiched chips under a bright field microscope with 4× or 10× objective lens for spot attachment and uniform cell printing. ***Note: Make sure to avoid water evaporation during chip inspection.***
17. Change growth medium every 1–2 days by printing fresh growth medium in the microwell chip, discarding the microwell chip containing old growth medium, and sandwiching the freshly prepared microwell chip.

4.5.3.6 Cell Printing in a Mixture of Alginate and Fibrinogen for 3D Cultures

Fibrinogen is a glycoprotein, which is converted into fibrin gel via polymerization in the presence of thrombin, and can be used for cell encapsulation. Fibrin gel is widely used in gene delivery, cell growth and differentiation, and tissue engineering applications [13–16]. Since it is a soft gel and susceptible to hydrolytic degradation by proteases, fibrinogen alone may not be suitable for microarray 3D bioprinting applications. However, fibrinogen can be mixed with alginate to improve physical strength and used to facilitate 3D cell cultures. In this chapter, we provide an experimental procedure for cell encapsulation and culture in a mixture of alginate, fibrinogen, and cells on the micropillar chip in the presence of thrombin [12].

1. Coat the micropillar chip with 2 mL of 0.01 % (w/v) PMA-OD in ethanol and dry it at room temperature at least for 2–3 h.
2. Prepare a mixture of 0.0033 % PLL and 25 mM $CaCl_2$ by mixing 0.01 % PLL with 37.5 mM $CaCl_2$ in distilled water at 1:2 ratio.
3. Print 60 nL of the PLL-$CaCl_2$ mixture on the micropillar chip using a work file (60 nL PLL_$CaCl_2$) and dry it for at least 2–3 h.
4. Take out one aliquot of 100 U/mL thrombin from the −20 °C freezer and mix it with growth medium at 1:10 ratio to achieve a final concentration of 10 U/mL thrombin. ***Note: Do not vortex it. Just mix it with a 1 mL pipette.***
5. Print 950 nL of growth medium containing 10 U/mL thrombin in the microwell chip placed on the chilling chip deck at 4–12 °C using a work file (medium_6 tip_6 block).
6. Immediately after printing, incubate the microwell chips with growth medium in a humid chamber with water to avoid water evaporation (Fig. 4.7).
7. Prepare 100 mg/mL fibrinogen freshly by adding 100 mg of fibrinogen in 1 mL of 0.9 % NaCl and shaking the solution slightly to dissolve. ***Note: Fibrinogen solution should not be vortexed. Fibrinogen solution cannot be stored, and should be prepared fresh every time.***

8. Prepare 1–4 million cells/mL in 0.75 % alginate containing 25 mg/mL fibrinogen by mixing 1 mL of 2–8 million cells/mL, 0.5 mL of 3 % alginate in distilled water, and 0.5 mL of 100 mg/mL fibrinogen using a 1 mL pipette. ***Note: Do not vortex the mixture as mammalian cells are fragile. Mix well by aspirating and dispensing the solution repeatedly (20 times) with a 1 mL pipette.***
9. Place the micropillar chips with dried PLL-$CaCl_2$ spots on the chilling chip deck at 4–12 °C. ***Note: Do not leave the micropillar chips too long on the chilling chip deck unattended as there might be severe water condensation observed on the chip surface, which will affect spot attachment.***
10. Add 100–200 μL/96-well of 1–4 million cells/mL in 0.75 % alginate containing 25 mg/mL fibrinogen in a 96-well plate. ***Note: Always use the same kind of the 96-well plates to avoid tip crashing due to Z height difference in 96-wells.***
11. Immediately before cell printing, resuspend the cells in 0.75 % alginate containing 25 mg/mL fibrinogen with a pipette. ***Note: For uniform cell printing and preventing solenoid valve and tip clogging, it is extremely important to resuspend the cells immediately before printing using a 100 μL pipette.***
12. Print 60 nL of the cell suspension in 0.75 % alginate containing 25 mg/mL fibrinogen on the micropillar chips placed on the chilling chip deck at 4–12 °C using a work file (Cell_Alginate_6 tip_6 block_60nL).
13. After 1–2 min gelation on the chilling chip deck, sandwich the micropillar chip with encapsulated cells with the microwell chip containing 950 nL of growth medium supplemented with 10 U/mL thrombin.
14. Place the sandwiched chips with the micropillar chip on the top in a humid cell incubation chamber or a petri dish with 10 mL water (Fig. 4.8B) and incubate the chips overnight in a 5 % CO_2 incubator at 37 °C for complete gelation of fibrinogen. ***Note: Unless cytotoxicity noticed due to thrombin, there will be no need to rinse the micropillar chip with growth medium.***
15. If necessary, inspect the sandwiched chips under a bright field microscope with 4× or 10× objective lens for spot attachment and uniform cell printing. ***Note: Make sure to avoid water evaporation during chip inspection.***
16. Change growth medium every 1–2 days by printing fresh growth medium (without thrombin) in the microwell chip, discarding the microwell chip containing old growth medium, and sandwiching the freshly prepared microwell chip.

4.5.3.7 Cell Printing in PuraMatrix for 3D Cultures

PuraMatrix from BD Biosciences is a synthetic peptide-based hydrogel commonly used for 3D cell cultures. The peptide component of PuraMatrix forms 3D fibrous structures with salts via self-assembly. The average pore size of the PuraMatrix hydrogel is 50–200 nm. One of the biggest problems of using PuraMatrix for 3D cell culture is its strong acidity, which can be toxic to many cells. Therefore, rinsing cells with growth media immediately after cell printing is of utmost importance.

1. Coat the micropillar chip with 2 mL of 0.01 % (w/v) PMA-OD in ethanol and dry it at room temperature at least for 2–3 h.
2. Sonicate 500 μL of 1 % PuraMatrix (from BD Biosciences) in a 1.5 mL Eppendorf tube for 30 min every time to decrease the viscosity. ***Note: If air bubbles are present, centrifuge the PuraMatrix stock solution.***
3. Prepare 0.25 % PuraMatrix by mixing 1 % PuraMatrix with sterile deionized water in a 1:3 (v/v) ratio.
4. Print 60 nL of 0.25 % PuraMatrix on the micropillar chip using a work file (60 nL PLL_$CaCl_2$) and dry it for at least 4 h. ***Note: If the micropillar chips with dried PuraMatrix are not being used within 24 h of the print, store overnight at 4 °C.***
5. For Hep3B cells, prepare RPMI medium containing 10 % FBS, 1 % Pen/Strep (100 x dilution), and 50 μg/mL gentamycin (1000 × dilution).
6. Print 950 nL of growth medium onto the microwell chip using a work file (medium_6 tip_6 block) while maintaining the chilling chip deck at 4–12 °C and incubate it in a humid chamber at 37 °C until use to avoid water evaporation. ***Note: Three microwell chips with growth medium are necessary for every micropillar chip with cells encapsulated in PuraMatrix.***
7. Mix PuraMatrix stock with 20 % sucrose in sterile deionized water in a 1:1 (v/v) ratio to generate a 0.5 % PuraMatrix solution in 10 % sucrose.
8. Trypsinize Hep3B cells, mix them with 7 mL growth media, and spin down cell suspension at 1200 rpm (200 g) for 4 min.
9. Remove the supernatant and resuspend cell pellet with 3 mL of growth media.
10. Take a small aliquot and measure cell density using a cell counter. ***Note: Sucrose might be toxic to cells, and interferes cell counting. Thus, do not use solutions containing sucrose yet.***
11. Seed cells in a T75 flask for near future experiments, if necessary.
12. Recentrifuge the cell suspension for the same speed and duration as described in step 8, completely remove growth media from cell pellet to eliminate salts, and resuspend the cells with 7 mL of 10 % sucrose in sterile deionized water to maintain osmolarity. ***Note: Resuspension of the cells must be done very gently with the 10 % sucrose solution as sucrose may lyse the cells.***
13. Centrifuge the cells at 200 g for 7 min, remove the supernatant, and then resuspend the cells in 10 % sucrose at 2× the final desired cell concentration (12 million cells/mL). ***Note: The gelation of PuraMatrix is initiated by salt concentrations greater than 1 mM in buffers or growth media. Therefore, do not combine PuraMatrix with salt-containing buffers or growth media until gelation is desired.***
14. Mix the 0.5 % PuraMatrix solution in 10 % sucrose with the 2× cell suspension in 10 % sucrose in a 1:1 (v/v) ratio very gently for a final concentration of 6 million cells/mL in 0.25 % PuraMatrix and 10 % sucrose. ***Note: Vigorous mixing can lyse the cells and release intracellular salts, resulting in premature gelation of PuraMatrix.***
15. Immediately after preparing the mixture freshly, print 60 nL of the cell suspension in 0.25 % PuraMatrix and 10 % sucrose on the micropillar chip with dried

PuraMatrix using a work file (60 nL cell_alginate_printing) while maintaining the chip deck at 4–12 °C. ***Note: This is one of the most important steps. Since PuraMatrix exhibits a pH of 2–3, printing cells within 5–10 min of mixing the cells and the PuraMatrix minimizes cell lysis and the chance for premature gelation. Freshly mix the 0.5 % PuraMatrix solution in 10 % sucrose with the 2× cell suspension in 10 % sucrose right before each cell printing.***

16. Wait two min for the gels to set.
17. Sandwich the micropillar chip with cells with the microwell chip containing 950 nL of pre-warmed growth medium for 30 min for gelation. Afterwards, discard the microwell chip. ***Note: Fresh growth medium without sucrose is used to remove 10 % sucrose and equilibrate the pH.***
18. Repeat the previous step once. ***Note: Since PuraMatrix is very acidic and toxic to cells, double rinsing with fresh growth medium is necessary every 30 min.***
19. Sandwich the micropillar chip again with another microwell chip containing pre-warmed media. Incubate the sandwiched chips in a humid incubation chamber placed in a 5 % CO_2 incubator at 37 °C for 3D cell cultures.

4.5.4 Virus Printing

Recombinant viruses have been used to transfer genes to human cells encapsulated in hydrogels on the micropillar chip. For example, recombinant lentiviruses with genes for neural stem cell (NSC)-specific biomarkers, including Synapsin1 for neuron differentiation, glial fibrillary acidic protein (GFAP) for astrocyte differentiation, myelin basic protein (MBP) for oligodendrocyte differentiation, and SOX2 for self-renewal, have been used on the chip platform to monitor real-time NSC differentiation. Recombinant adenoviruses carrying genes for cytochrome P450 (CYP450), including CYP1A2, CYP2C9, CYP2D6, CYP2E1, and CYP3A4, have been constructed and used on the chip platform for transiently controlled expression of CYP450 isoforms for metabolism-induced toxicity assays. Protocols for printing viruses are simple and straightforward, compared to printing cells and enzymes.

4.5.4.1 Measurement of Viral Titer in a 96-Well Plate

Prior to virus printing, it is important to know the number of virus particles in the solution. The multiplicity of infection (MOI) which is the number of virus particles per cell has to be measured to precisely control the expression levels of recombinant viruses used and minimize the basal toxicity due to excessive use of the viruses.

1. Seed HEK293 cells at a density of 10,000 cells in 100 μL Eagle's minimum essential medium (EMEM) per well in a 96-well plate.
2. Incubate the 96-well plate in a 5 % CO_2 incubator at 37 °C for 24 h. Check to see if cells are attached to the surface of 96 wells and spread out.

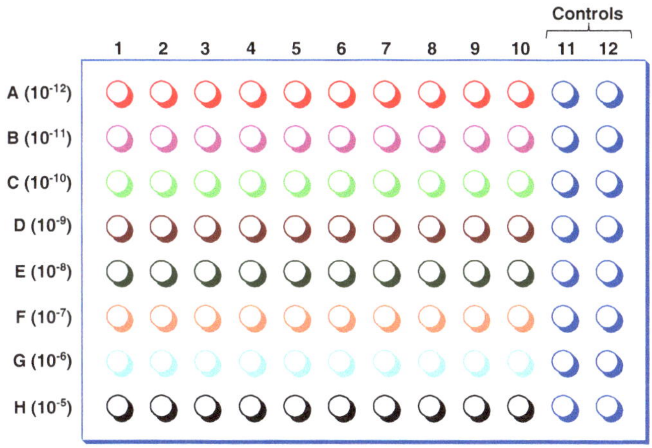

Fig. 4.10 Layout of a 96-well plate with different virus dilution in growth medium. The dilution factors used in the rows are recommended ranges. Sometimes there is need to increase/decrease the dilution factors accordingly. Growth medium alone (without viruses) is used as a control

3. Dilute viruses with EMEM at 10^{-5}–10^{-12} fold dilution (Fig. 4.10) and incubate the 96-well plate in the 5 % CO_2 incubator at 37 °C for 7–10 additional days.
4. After 7–10 days of incubation, check each well for cytopathic effect (CPE). This assay is only valid if the following are true.

    1) The control wells (columns 11 and 12) are completely CPE negative.
    2) The highest viral concentration (row H1–10) is completely CPE positive.
    3) The lowest viral concentration (row A1–10) is completely CPE negative.

5. Sum the fraction of wells in individual row that exhibit CPE.
6. Calculate viral titer by using the following equation:

$$Titer\left(\frac{pfu}{mL}\right) = 10^{0.8+n+\sum_{i=A}^{H} fi.}$$

Where A, B, C, …, and H are rows inspected
f = fraction of wells in row that are CPE positive
n = −1−log (highest concentration virus dilution)

4.5.4.2 Adenoviral Transduction on the Micropillar/Microwell Chip

1. Print 950 nL of recombinant adenoviruses diluted with growth medium at different MOIs into the microwell chip using a work file (medium_6tip_6block_950nL). The layout of the microwell chip containing viruses is shown in Fig. 4.11.
2. Sandwich the micropillar chip with 3D-cultured cells with the microwell chip containing adenoviruses and incubate the sandwiched chips in the humidified

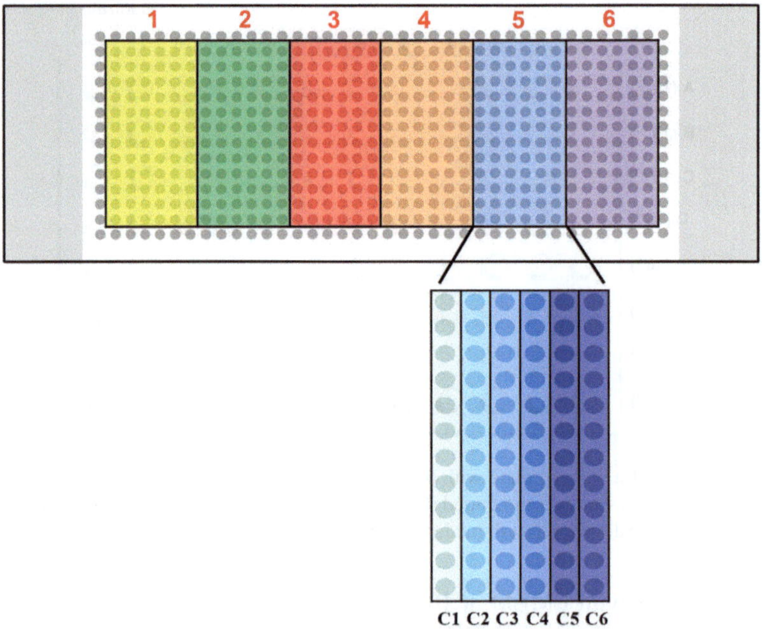

Fig. 4.11 Layout of the microwell chip containing different MOI of adenoviruses for gene expression. For example, C1 is no virus condition, C2 is 5 MOI of adenovirus carrying gene for GFP (Ad-GFP), C3 is 20 MOI of Ad-GFP, C4 is 10 MOI of Ad-GFP plus 10 MOI of Ad-RFP, C5 is 5 MOI of Ad-RFP, and C6 is 20 MOI of Ad-RFP

petri dish with 10 mL of distilled water placed in a 5 % CO_2 incubator at 37 °C for 24 h. ***Note: For 2D-grown cells, 4–6 h of virus exposure are commonly used. For 3D-cultured cells, it requires longer incubation time with viruses due to virus diffusion through the hydrogel matrix. Alginate is the worst hydrogel for virus infection because of strong binding affinity of virus particles to alginate. For higher efficiency of virus infection, 3D cells may be encapsulated in Matrigel, PuraMatrix, or alginate mixture with Matrigel and fibrinogen.***

3. During cell exposure to viruses, print 950 nL of growth medium (with no virus) onto the microwell chip using a work file (medium_6 tip_6 block) while maintaining the chilling chip deck at 4–12 °C and incubate it in a humid chamber at 37 °C until use to avoid water evaporation. ***Note: Two microwell chips with growth medium are necessary for every micropillar chip with infected cells for removing excess viruses.***
4. Discard the microwell chip containing viruses, and sandwich the micropillar chip with infected cells with the microwell chip containing 950 nL of fresh growth medium for 1 h for rinsing and removing excess viruses.
5. Discard the microwell chip again, and sandwich the micropillar chip with infected cells with the microwell chip containing fresh growth medium for gene expression.

6. After 48 h incubation, inspect the sandwiched chips under a fluorescent microscope with 4× or 10× objective lens for gene expression and spot attachment. ***Note: Make sure to avoid water evaporation during chip inspection.***
7. Dry the micropillar chip and observe fluorescence using the S+ scanner under green, red, and multiband filter settings, or measure activity of various drug metabolizing enzymes using appropriate kits.

4.6 Inspection of Cells Printed on the Micropillar Chips Using a Bright-Field Microscope

1. To inspect cells printed on the micropillar chip or in the microwell chip under a bright-field microscope (EVOS XL Core Imaging System, ThermoFisher Scientific), turn on the microscope by pressing the on/off switch on the right side of the microscope and wait until the starting messages disappear and the screen is ready for inspection (Fig. 4.12).
2. Select an objective lens for chip inspection. There are three different magnifications of objective lenses (including 4×, 10×, and 20×) installed in the bright-field microscope. ***Note: Typically, 4× objective lens is used for rapid inspection of spot detachment, and 20× objective lens is used for inspecting cell morphology.***
3. Place the chip with cells on the chip deck and adjust focus positions by rotating the coarse and fine focus adjustment knob for visualizing cells (Fig. 4.13a). ***Note: Make sure to avoid water evaporation during chip inspection.***
4. Move around the chip with cells to inspect spot detachment and uniform cells printing in each block. ***Note: Make sure that the overall number of cells in***

Fig. 4.12 Picture of the microscope with the on/off switch, objective lenses, and the focus adjustment knob: (*Left*) front view and (*Right*) side view

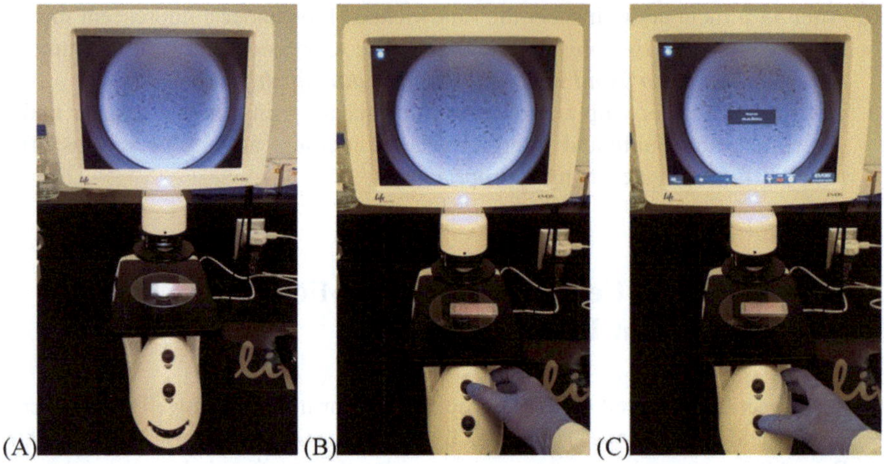

Fig. 4.13 (**a**) Adjusting focus positions, (**b**) freezing the image, and (**c**) saving the cell image on the chip.

spots printed by different solenoid valves are similar. The block-to-block variation (i.e., solenoid valve-to-solenoid valve variation) should be less than 20 %. In addition, spot detachment should be below 1 %.

5. Press the "Freeze" button to freeze the cell image before taking a picture (Fig. 4.13b).
6. Press the "Save" button to store the picture in the USB connected to the microscope (Fig. 4.13c). ***Note: Do not move the chip until the save message disappears from the screen.***

4.7 Coefficient of Variation (CV) and Z' Factor for Assay Validation

For robust assay development on the chip platform, it is important to measure the range of errors and chip-to-chip and day-to-day reproducibility. The coefficient of variation (CV) and the Z' factor are commonly measured to evaluate error ranges and robustness of an assay. The CV is defined as the ratio of the standard deviation (σ) to the average (μ).

$$CV = \frac{\sigma}{\mu}$$

It is the extent of variability in relation to the average of the signal, thus the inverse of the signal-to-noise ratio. The range of acceptable CV is typically below 20 %.

Instead of using signal to noise (S/N) or signal to background (S/B), the Z' factor is the most widely used to measure the robustness of a new assay. The Z' factor is defined as the following equation where the averages of maximum/minimum signals from positive/negative controls are μ_+/μ_- and the standard deviations of positive/negative controls are σ_+/σ_-.

$$Z' = 1 - \frac{3(\sigma_+ + \sigma_-)}{(\mu_+ - \mu_-)}$$

Assays with a Z' factor between 0.5 and 1.0 are considered excellent. In addition, assays with a Z' factor between 0 and 0.5 are considered marginal. The Z' factor is used to identify the quality of a new assay prior to testing a large numbers of compounds using high-throughput screening (HTS).

4.8 Summary

In this chapter, we have reviewed the essential methods for printing biological samples. These methods were optimized for each individual sample, as a suspension of cells in media will have a different viscosity than cells suspended in each of the various hydrogels. Additionally, we have optimized preparation and printing protocols for hydrogels with various gelation mechanisms, including ion-responsive, pH-responsive, enzymatically-activated, and thermos-responsive hydrogels. Optimizing these protocols are essential for ensuring cell viability and reproducing mechanical and physical interactions observed between cells and ECM *in vivo*. Additionally, with successful enzyme, virus, and compound printing protocols established, it is possible to transiently control gene expression, and predict drug efficacy and toxicity for current and future therapeutics.

References

1. Lee, D. W., Lee, M. Y., Ku, B., Yi, S. H., Ryu, J. H., Jeon, R., et al. (2014). Application of the DataChip/MetaChip technology for the evaluation of ajoene toxicity in vitro. *Archives of Toxicology, 88*(2), 283–290. doi:10.1007/s00204-013-1102-9.
2. Lee, M.-Y., Kumar, R. A., Sukumaran, S. M., Hogg, M. G., Clark, D. S., & Dordick, J. S. (2008). Three-dimensional cellular microarray for high-throughput toxicology assays. *Proceedings of the National Academy of Sciences, 105*(1), 59–63. doi:10.1073/pnas.0708756105.
3. Lee, M.-Y., Park, C. B., Dordick, J. S., & Clark, D. S. (2005). Metabolizing enzyme toxicology assay chip (MetaChip) for high-throughput microscale toxicity analyses. *Proceedings of the National Academy of Sciences, 102*(4), 983–987. doi:10.1073/pnas.0406755102.

4. Kadletz, L., Heiduschka, G., Domayer, J., Schmid, R., Enzenhofer, E., & Thurnher, D. (2015). Evaluation of spheroid head and neck squamous cell carcinoma cell models in comparison to monolayer cultures. *Oncology Letters*, 1281–1286. doi:10.3892/ol.2015.3487.
5. Pawar, S. N., & Edgar, K. J. (2012). Alginate derivatization: A review of chemistry, properties and applications. *Biomaterials, 33*(11), 3279–3305. doi:10.1016/j.biomaterials.2012.01.007.
6. Jeon, O., Bouhadir, K. H., Mansour, J. M., & Alsberg, E. (2009). Photocrosslinked alginate hydrogels with tunable biodegradation rates and mechanical properties. *Biomaterials, 30*(14), 2724–2734. doi:10.1016/j.biomaterials.2009.01.034.
7. Jeon, O., Powell, C., Ahmed, S. M., & Alsberg, E. (2010). Biodegradable, photocrosslinked alginate hydrogels with independently tailorable physical properties and cell adhesivity. *Tissue Engineering. Part A, 16*(9), 2915–2925. doi:10.1089/ten.tea.2010.0096.
8. Hsiong, S. X., Boontheekul, T., Huebsch, N., & Mooney, D. J. (2009). Cyclic arginine-glycine-aspartate peptides enhance three-dimensional stem cell osteogenic differentiation. *Tissue Engineering. Part A, 15*(2), 263–272. doi:10.1089/ten.tea.2007.0411.
9. Morritt, A. N., Bortolotto, S. K., Dilley, R. J., Han, X., Kompa, A. R., McCombe, D., et al. (2007). Cardiac tissue engineering in an in vivo vascularized chamber. *Circulation, 115*(3), 353–360. doi:10.1161/CIRCULATIONAHA.106.657379.
10. Ponce, M. L. (2009). Tube formation: An in vitro matrigel angiogenesis assay. *Methods in Molecular Biology, 467*, 183–188.
11. Thiele, J., Ma, Y., Bruekers, S. M. C., Ma, S., & Huck, W. T. S. (2014). 25th anniversary article: Designer hydrogels for cell cultures: A materials selection guide. *Advanced Materials, 26*(1), 125–148. doi:10.1002/adma.201302958.
12. Datar, A., Joshi, P., & Lee, M. Y. (2015). Biocompatible hydrogels for microarray cell printing and encapsulation. *Biosensors, 5*(4), 647–663. doi:10.3390/bios5040647.
13. Zhang, Z., He, Q., Deng, W., Chen, Q., Hu, X., Gong, A., et al. (2015). Nasal ectomesenchymal stem cells: Multi-lineage differentiation and transformation effects on fibrin gels. *Biomaterials, 49*, 57–67. doi:10.1016/j.biomaterials.2015.01.057.
14. Eyrich, D., Brandl, F., Appel, B., Wiese, H., Maier, G., Wenzel, M., et al. (2007). Long-term stable fibrin gels for cartilage engineering. *Biomaterials, 28*(1), 55–65. doi:10.1016/j.biomaterials.2006.08.027.
15. Luyckx, V., Dolmans, M.-M., Vanacker, J., Scalercio, S. R., Donnez, J., & Amorim, C. A. (2013). First step in developing a 3D biodegradable fibrin scaffold for an artificial ovary. *Journal of Ovarian Research., 6*(1). doi:10.1186/1757-2215-6-83.
16. Huang, Y.-C. Y., Dennis, R. R. G. R. R. G., Larkin, L., & Baar, K. (2005). Rapid formation of functional muscle in vitro using fibrin gels. *Journal of Applied Physiology, 98*(2), 706–713. doi:10.1152/japplphysiol.00273.2004.

Chapter 5
High-Content Cell Staining

Kyeong-Nam Yu, Pranav Joshi, and Moo-Yeal Lee

Contents

5.1 Introduction ... 105
5.1.1 Fluorescent Dyes ... 106
5.1.2 Immunofluorescence (IF) Assays with Antibodies ... 107
5.1.3 Fluorescent Proteins ... 108
5.2 Materials ... 109
5.2.1 Reagents for Fluorescence Staining ... 109
5.2.2 Reagents for Immunofluorescence Staining ... 109
5.2.3 Devices for Cell Staining ... 110
5.2.4 Preparation of Dye Stock Solutions in DMSO ... 111
5.3 Protocols ... 112
5.3.1 Staining Cells with Fluorescent Dyes ... 112
5.3.1.1 Preparation of a Saline Solution ... 112
5.3.1.2 Staining Cells on the Micropillar/Microwell Chip Platform with Fluorescent Dyes ... 112
5.3.2 Staining Cells on the Chip Platform with Fluorophore-Labeled Antibodies ... 116
5.3.2.1 Cell Fixation ... 117
5.3.2.2 Permeabilization of Cell Membranes ... 117
5.3.2.3 Blocking of Nonspecific Binding and Incubation with Primary/ Secondary Antibodies for Fluorescence Labeling and Detection ... 118
5.3.3 Measuring the Expression Levels of Drug Metabolizing Enzymes on a Chip ... 119
5.3.3.1 Tyramide Signal Amplification Kit (Life Technologies) ... 121
5.4 Summary ... 122
References ... 122

5.1 Introduction

Fluorescence-based cell imaging is an important technology for analyzing various biological processes at cellular and molecular levels [1]. Morphological and functional features in a cell can be labeled with multiple fluorescent probes/reagents, imaged with automated fluorescence microscopes, and quantified with image

M.-Y. Lee, *Microarray Bioprinting Technology*,
DOI 10.1007/978-3-319-46805-1_5

analysis algorithms [2]. In particular, high-content imaging (HCI) has gained popularity for systematic and accurate evaluation of drug candidates [3, 4] because of its capability to assess specific signals, including changes in the nucleus, organelle structure, protein translocation, oxidative stress, apoptosis/necrosis, mitochondrial impairment, calcium homeostasis, morphology, and phenotype profiling as readouts [2, 4]. Three different types of cell labeling are typically used for HCI assays, which include fluorescent dyes for direct cell staining, antibodies for immunofluorescent labeling, and genetically expressed fluorescent proteins such as green fluorescent protein (GFP) [1, 5]. This chapter summarizes basic cell staining protocols with various fluorescent dyes and antibodies for HCI assays on the micropillar/microwell chip platform.

5.1.1 Fluorescent Dyes

Fluorescent dyes (also known as fluorophores) are generally polyaromatic hydrocarbons or heterocycles which function at distinct excitation and emission wavelengths to generate fluorescence [6, 7]. Fluorescent dyes are taken up by cells and concentrated in different organelles based on charge or molecular affinity. Many fluorescent dyes are available commercially for various cell-based assays and applications. For example, Hoechst 33342, 4′,6-diamidino-2-phenylindole (DAPI), and Draq5 are commonly used for the assessment of nucleus morphology and cell counting. Tetramethyl rhodamine methyl ester (TMRM) and MitoTracker are used for measuring mitochondrial impairment. Propidium iodide, calcein AM, and ethidium homodimer-1 are commonly used for measuring cell viability. TO-PRO-3 and BOBO-1 are used for assessing permeability of cell membrane, whereas YO-PRO-1 has been used for apoptotic cell death. H_2DCFDA and BODIPY 665/676 are commonly used for measuring the generation of reactive oxygen (O_2) species (ROS) [8–15].

Compared to immunofluorescence assays and fluorescent protein assays, fluorescent dye-based cell staining offers several clear advantages, such as ease of use, high sensitivity for fluorescence detection, and a wide range of selection for specific cellular mechanisms [7]. Among a myriad of commercially available fluorescent dyes, proper evaluation of several factors are required prior to cell staining. For example, it is necessary to avoid excitation/emission spectrum overlapping among different fluorescent probes to obtain clear separation of fluorescence signals when multiple dyes are used simultaneously. Another problem is that fluorescent dyes generally have limited photostability. Thus, stained cells have to be always stored in the dark as most fluorescent molecules may be photobleached, as repeated illumination on stained cells can cause continuous decrease in fluorescent signals. Fluorescent dye solutions have to be prepared freshly because exposing the solution to ambient air for a long time can reduce the fluorescence intensity as dissolved oxygen (O_2) may interact with fluorescent dyes.

As some fluorescent dyes are pH sensitive, it is required to use a pH neutral buffer solution when preparing the working solution. Precaution has to be taken for background fluorescence from samples and sample holders when acquiring cell images. Finally, fluorescence intensity is only linearly proportional to the concentration of fluorophore in diluted solutions [1, 16].

5.1.2 Immunofluorescence (IF) Assays with Antibodies

As very few fluorescent dyes can stain proteins expressed within cells or secreted by cells specifically, immunofluorescent labeling with antibodies has been developed to resolve this issue and visualize the distribution of target proteins in biological samples. Target proteins can interact with antibodies, which belong to a family of globular proteins called immunoglobulins (Igs) (Fig. 5.1). Ig has four polypeptide chains divided into two heavy (~50 kD, H) chains and two light (~23 kD, L) chains, one of which has a specific binding site for the epitope of an antigen (i.e., target protein). These specific binding of an antibody to an antigen is the major principle of IF assays [17].

Two types of antibodies such as monoclonal (mAbs) and polyclonal antibodies (pAbs) are commonly used for IF assays. The major difference is that a mAb recognizes one epitope on an antigen whereas a pAb recognizes multiple epitopes on any one antigen. Thus, a mAb is used to detect a specific target protein which leads to reproducible cell staining with significantly lower background fluorescence. On the other hand, a pAb is used to detect a target protein with low expression levels and amplify fluorescent signals due to multiple binding sites. The difference in their properties are summarized in Table 5.1 [17].

The major advantages of IF staining are their wide applications with sectioned tissues, primary cells, and cultured cell lines as well as their capability to detect proteins, glycans, and small biological molecules. The IF-labeled samples can be

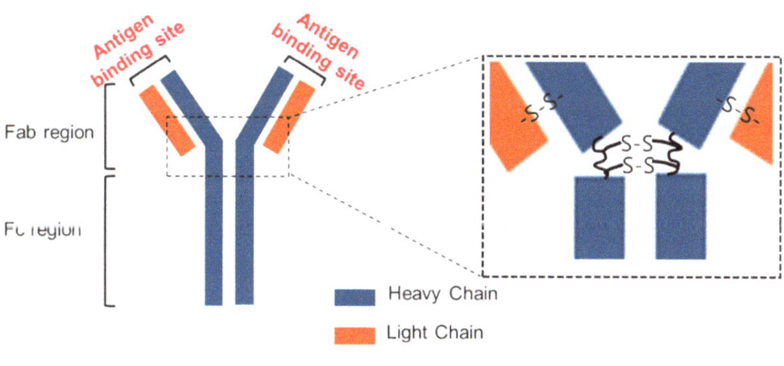

Fig. 5.1 Schematic diagram depicting the structure of an antibody

Table 5.1 Properties of monoclonal antibody (mAb) and polyclonal antibody (pAb)

mAb	pAb
• High specificity, detects only one epitope on an antigen	• High affinity, binds multiple epitopes on any one antigen and amplifying signals for robust detection for immunoprecipitation (IP) and chromatin immunoprecipitation (ChIP)
• Reduces background staining and highly reproducible between experiments	• High background and potentially generates inaccurate results
• Used as primary antibody	• Used as secondary antibody
• Expensive and takes a long time for hybridoma production	• Inexpensive and relatively easy to produce
• Too susceptible to the loss of epitope after chemical fixation	• High batch-to-batch variability

analyzed quantitatively and qualitatively by monitoring the fluorescent intensity. Diverse primary or secondary antibodies with fluorescent tags or enzymes have been developed and used for HCI assays in the fields of modern medicine, molecular biology, and toxicology. However, a variety of factors can interrupt the quality of IF assays, including nonspecific antibody binding, excitation/emission spectrum overlapping, autofluorescence, and photobleaching. For example, some mammalian cells contain flavin coenzymes such as flavin adenine dinucleotide (FAD) and flavin mononucleotide (FMN) (absorption: 450 nm and emission: 515 nm) and reduced pyridine nucleotide such as nicotinamide adenine dinucleotide (NADH) (absorption: 340 nm and emission: 460 nm) that may have spectra overlapping with fluorophores tagged on antibodies [18]. In addition, fluorophores bound to antibodies can react with ROS generated by biological reactions, resulting in destruction of the fluorophore [19]. Furthermore, aldehyde used for cell fixation can react with proteins and amines, generating unwanted fluorescence [20].

5.1.3 *Fluorescent Proteins*

Cells expressing fluorescent proteins are used in HCI assays to investigate various cellular processes, such as protein trafficking, gene activation, protein-protein interactions, organelle condition, and cellular development/differentiation [5]. Fluorescent proteins in mammalian cells are expressed by transferring genes encoding fluorescent proteins into the cells using plasmids, viral vectors, or other means. The processes known as transfection or transduction can be either transient or permanent. In transient transfection/transduction, the gene introduced into the cell does not integrate into the chromosomes, therefore expressing the protein for a short period of time. On the other hand, the gene can be permanently incorporated into the genome to create transformed cell lines. Selection of fluorescent proteins for HCI generally depends on several properties, including efficiency of protein expression, toxicity of the protein in the cells, brightness, photostability, and insensitivity to environmental effects [21]. For example, enhanced green fluorescent protein (EGFP) offers higher photostability than emerald in green spectrum ranges, and mOrange offers high brightness but low photostability. In addition, mCherry and mPlum offer high brightness and photostability in red and far-red spectrum ranges. Moreover, CyPet offers better photostability than mCFP, but its expression efficiency is relatively low at 37 °C [21]. The major advantage of fluorescent proteins is the possibility of fluorescence labeling to almost any protein of interest [5]. However, fluorescent proteins are relatively large in size, which can affect the function and localization of endogenous proteins of interest [5].

5.2 Materials

5.2.1 *Reagents for Fluorescence Staining*

- Hoechst 33342 (Thermo Fisher Scientific)
- Tetramethyl rhodamine methyl ester (TMRM; Thermo Fisher Scientific)
- Fluo-4 acetoxy methyl ester (Fluo-4 AM; Thermo Fisher Scientific)
- Monochlorobimane (mBCl; Thermo Fisher Scientific)
- Live/dead viability/cytotoxicity kit for mammalian cells (Thermo Fisher Scientific)
- YO-PRO-1 (Thermo Fisher Scientific)
- Propidium iodide (Thermo Fisher Scientific)

5.2.2 *Reagents for Immunofluorescence Staining*

- 1× Phosphate-buffered saline (PBS) containing 137 mM NaCl, 2.7 mM KCl, 10 mM Na_2HPO_4, and 2 mM KH_2PO_4. Adjust pH to 7.4 with 1 M HCl.
- 1× Tris-buffered saline (TBS) containing 50 mM Tris-HCl and 150 mM NaCl. Adjust pH to 7.4 with 1 M HCl. ***Note: If cells to be stained are encapsulated in alginate, TBS should only be used in the staining process, instead of PBS. PBS is incompatible with alginate samples. TBS is used to prevent degradation of alginate by chelating agents such as phosphate ions and EDTA.***
- Paraformaldehyde (Sigma-Aldrich)
- Triton X-100 (Sigma-Aldrich)
- Bovine serum albumin (BSA) (Sigma-Aldrich)
- Antibodies
  - Primary antibodies should be selected according to target proteins. ***Note: The primary antibody should be considered based on the species of the target protein stained, which avoids cross-reactivity of the secondary antibody with endogenous antibodies in the sample. For example, if a mouse protein is studied, select a primary antibody obtained from different species (e.g., human, rabbit, or goat).***

Table 5.2 Considerations to select secondary antibodies [22]

Considerations	Description
Reactivity	Anti-mouse, anti-rat, anti-chicken, anti-rabbit, anti-goat, anti-human, *etc.*
Non fluorescent conjugates	Horseradish peroxidase (HRP)-labeled, alkaline phosphatase (AP)-labeled, biotin, unconjugated, *etc.*
Fluorescent conjugates	Alexa fluor™ dyes, fluorescein isothiocyanate (FITC), tetramethylrhodamine (TRITC), DyLight™ dyes, rhodamine, Texas Red™ & Texas Red™-X, R-phycoerythrin (R-PE), allophycocyanin (APC), Qdot™ probes, Pacific dyes, *etc.*
Target Ig class	IgG, IgM, IgA, IgG2a, IgG1, IgD, Kappa or lambda type light chain, *etc.*

- ○ Secondary antibodies should be selected based on reactivity against primary antibodies, fluorescent conjugates, and target immunoglobulin (Ig) class (recognition of whole IgG or any fragments) (Table. 5.2).
  - Host species: If the primary antibody is from a mouse, the secondary antibody used should be anti-mouse.
  - Experimental methods: For immuno-staining assays (e.g., IF, IHC, *etc.*), it requires to select a secondary antibody conjugated with fluorescent dyes or enzymes.
  - Class/subclass of antibody: In general, polyclonal primary antibodies are produced from rabbit, goat, or sheep, and the majority of them has diverse Ig isotypes. Therefore, secondary antibodies used should be anti-Ig antibodies.
  - Ig class/subclass and type/subtype: Ig has diverse isoforms, and each Ig isoform has a unique sequence for antigen binding. Before selecting a secondary antibody, the Ig class has to be checked.

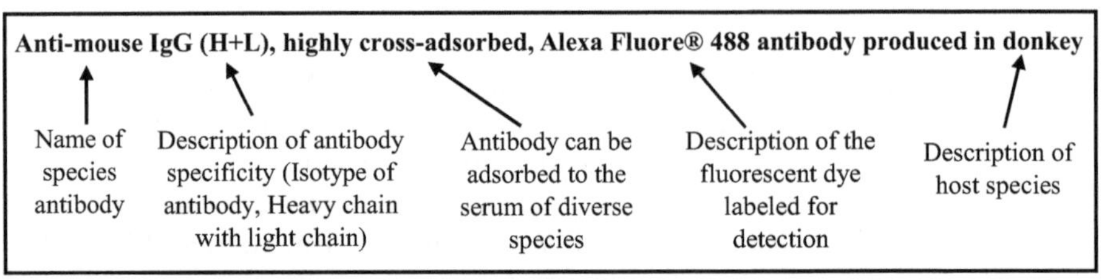

- 4′,6-Diamidino-2-phenylindole (DAPI; Sigma-Aldrich)
- Mounting solution with DAPI staining (Sigma-Aldrich)
- Mounting solution without DAPI staining (Fisher Scientific)

5.2.3 Devices for Cell Staining

- Deep-well staining plate (Samsung Electro-Mechanics, Co. or SEMCO, Suwon, South Korea)
- Shallow-well staining plate (SEMCO, Suwon, South Korea)

5.2.4 Preparation of Dye Stock Solutions in DMSO

- **Hoechst 33342**
  - Dissolve 20 mg of Hoechst 33342 in 3.25 mL DMSO to prepare a stock solution of 10 mM Hoechst 33342.
  - Store aliquots of 10 mM Hoechst 33342 in a −20 °C freezer protected from light until use.
  - To prepare a working solution of 25 μM Hoechst 33342, add 20 μL of the stock solution in 8 mL of Dulbecco's phosphate buffered saline (DPBS). ***Note: If possible, do not reuse the stock solution of the dye for reproducible results.***
- **Tetramethyl rhodamine methyl ester (TMRM)**
  - Dissolve 25 mg TMRM in 1 mL DMSO and vortex it for 1 min to prepare a stock solution of 50 mM TMRM.
  - Store aliquots of 50 mM TMRM in a −20 °C freezer protected from light until use.
  - Dilute the stock solution 100-fold to get a working stock concentration of 0.5 mM. Store aliquots of 0.5 mM TMRM in a −20 °C freezer, and do not reuse this stock solution for reproducible results.
  - To prepare a working solution of 0.5 μM TMRM, add 8 μL stock in 8 mL DPBS.
- **Fluo-4 acetoxy methyl ester (Fluo-4 AM)**
  - Dissolve 50 μg of Fluo-4 AM in 23 μL DMSO to prepare a stock solution of 2 mM Fluo-4 AM. ***Note: Fluo-4 AM may require addition of 20 % (w/v) Pluronic F-127 to enhance solubility.***
  - Store aliquots of 2 mM Fluo-4 AM in a −20 °C freezer protected from light until use.
  - To prepare a working solution of 5 μM Fluo-4 AM, add 10 μL stock in 2 mL DPBS.
- **Monochlorobimane (mBCl)**
  - Dissolve 25 mg of mBCl in 550 μL DMSO to prepare a stock solution of 200 mM mBCl.
  - Store aliquots of 200 mM mBCl in a −20 °C freezer protected from light until use.
  - To prepare a working solution of 100 μM mBCl, add 4 μL stock in 8 mL DPBS.
- **Calcein AM**
  - Calcein AM comes in a liquid form in the live/dead viability/cytotoxicity kit at a concentration of 4 mM without the need for adding solvent.
  - Store aliquots of calcein AM in a −20 °C freezer protected from light until use.
  - To prepare a working solution of 1 μM calcein AM, add 2 μL stock in 8 mL DPBS.

- **Ethidium homodimer-1**
  - Ethidium homodimer-1 comes in a liquid form in the live/dead viability/cytotoxicity kit at a concentration of 2 mM without the need for adding solvent.
  - Store aliquots of ethidium homodimer-1 in a −20 °C freezer protected from light until use.
  - To prepare a working solution of 1 μM ethidium homodimer-1, add 4 μL stock in 8 mL DPBS.
- **Propidium iodide**
  - Propidium iodide comes in a liquid form at a concentration of 1.5 mM without the need for adding solvent.
  - Store aliquots of 1.5 mM propidium iodide in a 4 °C refrigerator protected from light until use. ***Note: Propidium iodide is stable for at least 6 months when stored at 4 °C.***
  - To prepare a working solution of 5 μM propidium iodide, add 27 μL stock in 8 mL DPBS.
- **YO-PRO-1**
  - YO-PRO-1 comes in liquid form at a concentration of 1 mM without the need for adding solvent.
  - Store aliquots of 1 mM YO-PRO-1 in a −20 °C freezer protected from light until use.
  - To prepare a working solution of 10 μM YO-PRO-1, add 80 μL stock in 8 mL DPBS.

5.3 Protocols

5.3.1 Staining Cells with Fluorescent Dyes

5.3.1.1 Preparation of a Saline Solution

1. Dissolve 8.1 g of NaCl and 2.9 g of $CaCl_2 \cdot H_2O$ in 1 L of sterile deionized water to prepare a saline solution containing 140 mM NaCl and 20 mM $CaCl_2$. ***Note: The saline solution with 20 mM $CaCl_2$ is used only if the viability of the cells encapsulated in alginate on the chip is not affected by the salts. Alginate is sensitive to chelating agents such as phosphate ions in some buffer solutions. DPBS is recommended when working with hydrogels other than alginate.***

5.3.1.2 Staining Cells on the Micropillar/Microwell Chip Platform with Fluorescent Dyes

1. Prepare working solutions of desired fluorescent dyes in the saline solution prior to initial washing of the chip, and keep it protected from light (Table 5.3).

Table 5.3 The working conditions of commonly used fluorescent dyes

Fluorescent dye	Mol. wt.	Quantity (mg)	Stock conc. (mM)	Solvent added (μL)	Solvents used to dissolve dyes	Working stock conc. (mM)	Working conc. (μM)	Working solvent	Dye volume (μL)	Working volume (mL)
Hoechst 33342	615.99	20	10	3250	DMSO	10	25	DPBS	20	8
Tetramethyl rhodamine methyl ester (TMRM)	500.93	25	50	998	DMSO	0.5	0.5	DPBS	8	8
Fluo-4 acetoxy methyl ester (Fluo-4 AM)	1096.95	0.05	2	23	DMSO	1	5	DPBS	10	2
YO-PRO-1	629.32	1 mL	1	NA	DMSO	1	10	DPBS	80	8
Propidium iodide (PI)	668.4	10 mL	1.5	NA	Water	1.5	5	DPBS	27	8
Monochlorobimane (mBCl)	226.66	25	200	551	DMSO	200	100	DPBS	4	8
BODIPY 665/676	448.32	5	10	1115	DMSO	10	2.5	DPBS	2	8
Calcein AM	994.87	40 μL	4	NA	DMSO	4	1	DPBS	2	8
Ethidium homodimer-1	856.77	200 μL	2	NA	DMSO	2	1	DPBS	4	8

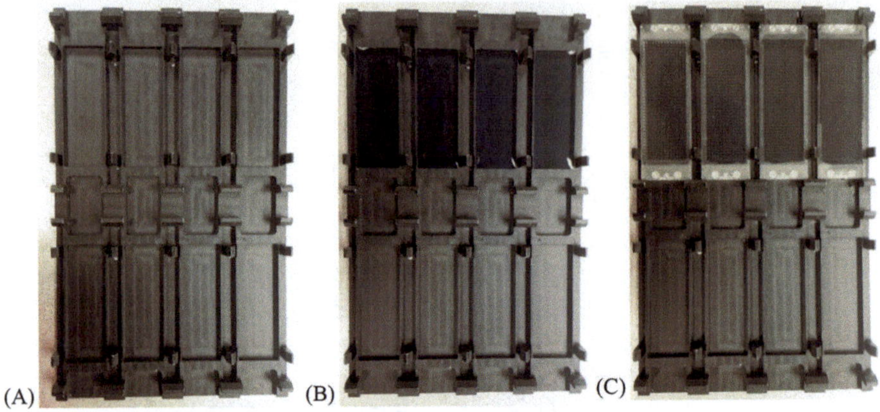

Fig. 5.2 Pictures of (**a**) an empty deep-well plate, (**b**) a deep-well plate with a saline solution and (**c**) a deep-well plate with the micropillar chips placed on the saline solution for cell rinsing.

2. Using 10 mL serological pipette, aspirate the saline solution and dispense 5.5 mL in each well of the deep-well staining plate (Fig. 5.2B). ***Note: Make sure to avoid dust particles while washing and staining. To remove dust and tiny particles from deep-well and shallow-well plates, blow air on the wells of both deep-well and shallow-well staining plates before starting the cell staining process.***
3. Place the micropillar chip facing down on top of the deep well with 5.5 mL of the saline solution so that cells on the micropillars can be immersed in the solution, and rinse the chip twice for 5 min each (Fig. 5.2C). ***Note: The cells on the chip are immersed in the saline solution to remove growth media and fetal bovine serum (FBS), which may interfere cell staining. Make sure that there are no bubbles trapped underneath the chip.***
4. Add 2 mL of a desired fluorescent dye solution in each well of a shallow-well plate (Fig. 5.4B). ***Note: Spread the dye solution with a pipette tip evenly on the surface of the shallow well.*** As the surface of the shallow well is relatively hydrophobic, it may not be spread well enough for uniform cell staining.
5. Once the chip is rinsed on the deep-well plate, drain an excess saline solution from the chip by tilting the chip at an angle of 45° and remove the remaining saline solution from the side of the chip with a paper towel (Fig. 5.3). ***Note: Do not dry the cell spots on the micropillar chip as spot drying will cause cell death.***
6. Immediately after draining the excess saline solution, place the micropillar chip facing down on top of the shallow well with 2 mL of the desired fluorescent dye solution and incubate for 60 min for cell staining (Fig. 5.4C). ***Note: The staining time can be varied depending on the fluorescent dye used and the morphology of cells stained.***

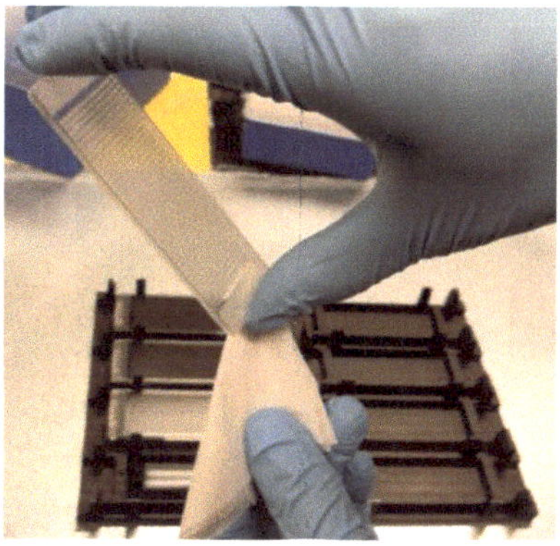

Fig. 5.3 Removing the excess saline solution from the side of the micropillar chip with a paper towel by tilting the chip at a 45° angle

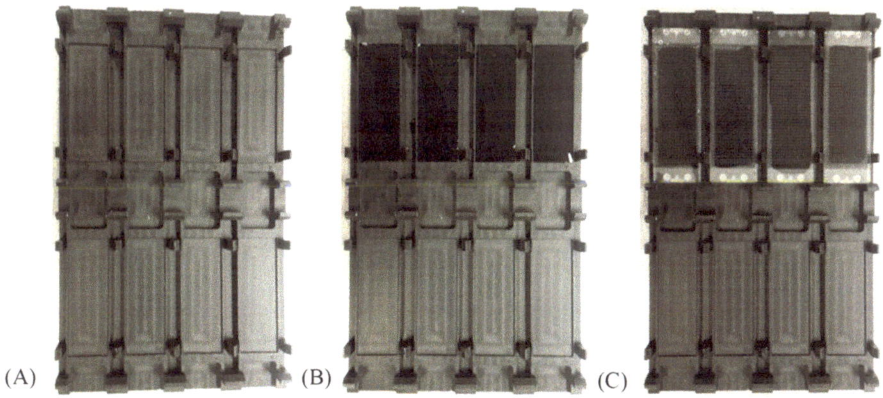

Fig. 5.4 Pictures of (**a**) an empty shallow-well plate, (**b**) a shallow-well plate with a fluorescent dye solution and (**c**) a shallow-well plate with the micropillar chips placed on the fluorescent dye solution for cell staining

7. During the period of cell staining, cover the entire shallow-well plate with the chips with aluminum foil. ***Note: The majority of fluorescent dyes are very sensitive to light illumination. To avoid photobleaching of fluorescent dyes during staining, make sure to prevent the chips from light exposure***.
8. After cell staining, remove an excess dye solution by rinsing the micropillar chip with stained cells twice for 10 min each in the deep-well plate with 5.5 mL of the saline solution. The micropillar chips should be protected from light during the rinsing steps too.
9. After rinsing, drain and remove the excess saline solution from the chip as described in Step 5.

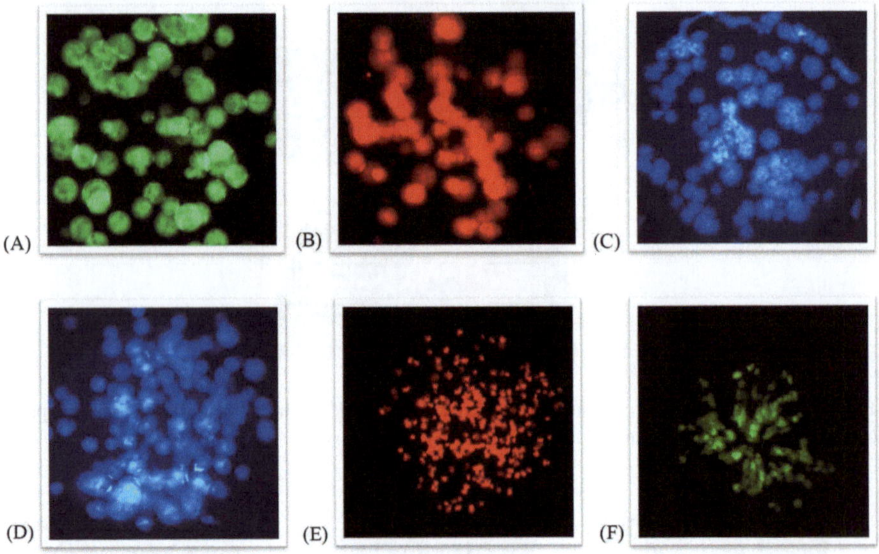

Fig. 5.5 Representative images of Hep3B cells cultured in 3D (60 nL) on the micropillar chip and stained with (**b**) calcein AM for cell viability, (**b**) TMRM for mitochondrial membrane potential, (**c**) Hoechst 33342 for nucleus morphology and cell count, (**d**) mBCl for glutathione level, (**e**) propidium iodide for cell viability and necrosis, and (**f**) Fluo-4 AM for intracellular calcium level

10. Dry the stained chips completely in the dark for at least 2 h. Once the micropillar chips are completely dried, they are ready for scanning. As an example, images of 3D-cultured cells stained with various fluorescent dyes are presented in Fig. 5.5. ***Note: Do not scan a wet micropillar chip as dripping water with salts can damage objective lenses and filter sets in the S+ Scanner. Always completely dry the micropillar chip or seal the microwell chip with wet samples with a gas-permeable sealing membrane (Sigma-Aldrich) before scanning.***

5.3.2 Staining Cells on the Chip Platform with Fluorophore-Labeled Antibodies

In general, immunofluorescence (IF) assays on the chip consist of cell fixation, permeabilization of cell membranes, blocking of nonspecific binding, and incubation with primary/secondary antibodies for fluorescence labeling and detection. Carrying out IF assays on the micropillar/microwell chip require careful optimization of experimental conditions due to several reagents that can facilitate spot detachment. For example, some detergents used can detach cell spots from the micropillar chip and phosphate ions in PBS buffers can degrade alginate spots.

5.3.2.1 Cell Fixation

1. Rinse the cells by immersing the micropillar chip containing cells twice in 5.5 mL of 1× sterilized PBS (or 1× TBS) in the deep-well plate for 5 min each.
2. Incubate the chip in the shallow-well plate with 2 mL of a fixation reagent for 10–15 min. ***Note: Refer Table 5.4 for various fixation reagents and their methods.***
3. Rinse the cells three times in 5.5 mL of 1× sterilized PBS (or 1× TBS) in the deep-well plate for 3 min each.

5.3.2.2 Permeabilization of Cell Membranes

4. Incubate cells with permeabilization reagents for 10 min. ***Note: Refer Table 5.5 for various permeabilization reagents and their methods.***
5. After permeabilization, rinse cells with 1× PBS or 1× TBS.

Table 5.4 Fixation reagents commonly used for target antigens

Fixation reagents	Methods of use	Target antigens
Ice-cold acetone (100 %)	• Fix cells in −20 °C acetone for 5–10 min	Large proteins (e.g., immunoglobulin) [23, 24]
	• The chip made of polystyrene may not be compatible with	
	• 100 % acetone	
	• No permeabilization step needed after acetone fixation	
Ice-cold methanol (100 %)	• Fix cells in −20 °C methanol for 5–10 min	
	• No permeabilization step needed after methanol fixation	
4 % Paraformaldehyde with 1 % glutaraldehyde	• Fix cells in 4 % paraformaldehyde with 1 % glutaraldehyde for 10–20 min	Most proteins, peptides, and enzymes with low molecular weight [25]
	• Rinse cells briefly with 1× PBS (or 1× TBS)	
4 % Paraformaldehyde	• Fix cells in 4 % paraformaldehyde for 10–20 min	Most proteins, peptides, and enzymes with low molecular weight [26]
	• Rinse cells briefly with 1× PBS (or 1× TBS)	
10 % Neutral-buffered formalin (NBF)	• Fix cell in 10 % neutral buffered formalin for 10–20 min	Most proteins, peptides, and enzymes with low molecular weight [27]
	• Rinse cells briefly with 1× PBS (or 1× TBS)	
Carnoy's solution	• Carnoy's solution contains 60 % ethanol, 30 % chloroform, and 10 % acetic acid	Nucleic acids [28]
	• Fix cells in Carnoy's solution for 10–20 min and rinse the cells with 1× PBS (or 1× TBS)	

Table 5.5 Various permeabilization reagents and their methods of use

Permeabilization reagents		Method of use	References
Solvents	Acetone and methanol	• Acetone and methanol can be used to fix and permeabilize cells simultaneously • It can be used with crosslinking agents such as formaldehyde	[29]
Detergents	Triton X-100 and NP-40	• Treat cells with 0.1–0.2 % of the detergents in PBS for 10 min • It can partially dissolve the nuclear membrane, thus suitable for nuclear antigen staining	[30]
	Tween 20, saponin, digitonon (0.01 %), and leucoperm	• Treat cells with detergents for 10–30 min • These detergents are mild membrane solubilizers and make large pores on cellular membrane without dissolving membrane • It is suitable for cytoplasma membrane or soluble nuclear antigen staining	[30–32]

Table 5.6 Blocking reagents, their methods of use, and limitations

Blocking reagents	Methods of use	Characteristics	References
Normal serum	• Use it before introduction of primary antibody • This can be used for the dilution of antibodies	• Common blocking reagent • Make sure to use proper serum of species. It may interact with detection antibody	[33, 34]
Protein solution (0.1–5 % BSA, gelatin, skim milk)	• Use it before introduction of primary antibody • This can be used for the dilution of antibodies	• Can be made easily in labs • Must be made freshly before use • Competitively binding to analogous ligands	[35, 36]
Pre-formulated commercial buffers	• Use it before introduction of primary antibody • This can be used for the dilution of antibodies	• Contain a refined concentration of individual proteins • Using to diverse application and improved shelf life	[37]

5.3.2.3 Blocking of Nonspecific Binding and Incubation with Primary/Secondary Antibodies for Fluorescence Labeling and Detection

6. After cell fixation and permeabilization, incubate the cells in 1× PBS (or 1× TBS) containing 3 % BSA (or other blocking reagents) for 1 h to prevent non-specific binding of antibodies. ***Note: Blocking reagents may vary depending on the application. Refer to Table 5.6 for various blocking reagents and their methods of use.***

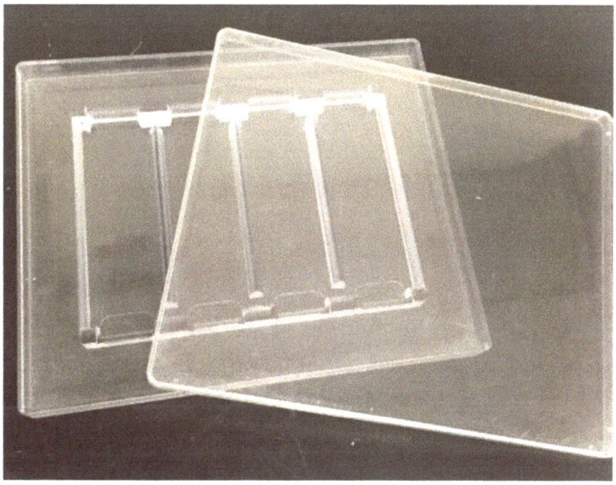

Fig. 5.6 Picture of a humidified chamber for cell staining with primary/secondary antibodies

7. Dilute a primary antibody in 1× PBS (1× TBS) containing a blocking reagent. The dilution factor of the primary antibody can be varied and determined experimentally. Check the manufacturer protocol or references. ***Note: A typical dilution factor ranges from 1:50 to 1:500.***
8. Incubate the cells overnight with the primary antibody at 4 °C in a humidified chamber (Fig. 5.6).
9. Decant the primary antibody and rinse the cells three times in 1× PBS containing Triton X-100 (1× PBS-T) or 1× TBS containing Triton X-100 (1× TBS-T) for 10 min each on a plate shaker.
10. Incubate the cells with a secondary antibody in 1× PBS (or 1× TBS) containing 3 % BSA for 1–3 h at room temperature in the humidified chamber. ***Note: For proper multicolor staining, ensure that primary antibodies and their corresponding secondary antibodies are from different species.***
11. Decant the secondary antibody and rinse three times in 1× PBS-T (or 1× TBS-T) for 10 min each in the dark on the plate shaker.
12. Stain the cells with DAPI at a final concentration of 500 nM for 10 min.
13. Rinse the cells twice with 1× sterile PBS-T (or 1× TBS-T) for 3 min each.
14. Apply a mounting solution for image acquisition.

5.3.3 Measuring the Expression Levels of Drug Metabolizing Enzymes on a Chip

1. Prepare 1× Tris-buffered saline (TBS) containing 0.1 % Tween 20 (1× TBS-T)
2. Rinse the cells by immersing the micropillar chip containing cells three times in 25 mL of 1× TBS in a petri dish (10 cm in diameter) for 5 min each. ***Note: Place the micropillar chip face down in the petri dish containing 1× TBS to allow the chip floating on the surface of the solution. Make sure not to damage cell spots***

Table 5.7 Examples of primary antibodies for detecting cytochrome P450 isoforms

Name	Target protein	Origin	Dilution factor	Company
Rabbit polyclonal anti-CYP1A1	CYP1A1	Rabbit	1:1000	Abcam
Rabbit polyclonal anti-CYP3A4	CYP3A4	Rabbit	1:100	Thermo Fisher Scientific
Rabbit polyclonal anti-CYP2C9	CYP2C9	Rabbit	1:100	Thermo Fisher Scientific

by touching the bottom of the petri dish when lifting the floating chip up with a pair of tweezers. Do not shake in order to avoid spot detachment.

3. Fix and permeate the cells by immersing the chip in 25 mL of 1× TBS containing 3.7 % formaldehyde for 20 min at room temperature and then transferring the chip in 25 mL of 1× TBS containing 0.15 % Triton X-100 for 10 min. In case the chip made of glass is used, fix and permeate the cells by immersing the glass chip in 25 mL of methanol and acetone (1:1, v/v) for 20 min at −20 °C.
4. Rinse the cells by immersing the chip twice in 25 mL of 1× TBS containing 0.15 TBS for 5 min each.
5. Dry the chip under weak nitrogen gas (or air) stream.
6. Incubate the chip overnight in 25 mL of a blocking buffer in TBS (SuperBlock from Fisher Scientific) at 4 °C.
7. Wash the chip three times in 25 mL of TBS-T for 5 min each. ***Note: Tween 20 in TBS-T is necessary to reduce hydrophobic interactions between protein and surface.***
8. Label the cells with a primary antibody (e.g., mouse anti-human IgG from Life Technologies) by diluting the primary antibody (typically 1:100–1:500, v/v, see primary antibody dilutions used in Table 5.7 as a reference) in TBS-T containing 1 % (w/v) bovine serum albumin (BSA), applying 2 mL of the primary antibody solution to the shallow-well plate, and then incubating the chip stamped onto the staining plate overnight at 4 °C.
9. Rinse the cells by immersing the chip three times in 25 mL of TBS-T for 15 min each with gentle shaking.
10. Label the cells with a secondary antibody (e.g., HRP-conjugated goat anti-mouse IgG from Life Technologies T20912) by diluting the secondary antibody (typically 1:500–1:1000, v/v) in TBS-T containing 1 % BSA, applying 2 mL of the secondary antibody solution to the shallow-well plate, and then incubating the chip stamped onto the staining plate for 3 h at room temperature.
11. Rinse the cells by immersing the chip three times in 25 mL of TBS-T for 15 min each with gentle shaking.
12. Label the cells with Alexa Fluor 488 tyramide (Tyramide signal amplification kit, Life Technologies T20912) by diluting the tyramide stock (1:200, v/v) in amplification buffer containing 0.0015 % H_2O_2 just prior to labeling, applying

2 mL of the tyramide working solution to the shallow-well plate, and incubating the chip stamped onto the staining plate for 30 min at room temperature. To freshly prepare the tyramide working solution, add 5 μL of 30 % H_2O_2 to 995 μL of amplification buffer and then 20 μL of this intermediate solution (0.15 % H_2O_2) to further a 1980 μL of amplification buffer.

13. Rinse the cells by immersing the chip three times in 25 mL of 1× TBS for 15 min each. ***Note: Tween 20 is unnecessary for the final rinsing.***
14. Dry the chip under weak air stream.
15. Scan the chip for green fluorescence using S+ Scanner. ***Note: The fluorescence signal of β-actin (determined by the same protocol using triplicate chips for each condition) can be used as an internal control.***

5.3.3.1 Tyramide Signal Amplification Kit (Life Technologies)

Contents

- Labeled tyramide (Component A), one vial
- Dimethylsulfoxide (DMSO; Component B), 200 μL
- HRP-conjugated secondary antibody (Component C), 100 μg
- Blocking reagent (Component D), 3 g
- Amplification buffer (Component E), 25 mL (containing 0.02 % thimerosal)
- Hydrogen peroxide (H_2O_2; Component F), 200 μL of a 30 % stabilized solution

Upon receipt and prior to use, the kit should be stored at −20 °C, desiccated and protected from light.

Preparation of Solutions

1. Prepare tyramide stock solution by dissolving the solid material provided (Component A) in 150 μL of DMSO (Component B). Invert the vial several times to dissolve any tyramide coating the sides of the vial. Store unused portions of this stock solution in small aliquots at −20 °C, desiccated and protected from light.
2. Prepare the HRP-conjugated antibody stock solution by reconstituting the material provided (Component C) in 200 μL of PBS. This solution may be stored at 2–6 °C for up to 3 months if required. Optionally, add 0.02 % thimerosal as a preservative. ***Note: Sodium azide must not be used for this purpose.***
3. Prepare a 1 % (10 mg/mL) solution of blocking reagent in PBS. ***Note: We recommend preparing only as much as is needed for immediate use. However, unused solution can be stored frozen at −20 °C for 1 month if necessary.***
4. Prepare amplification buffer/0.0015 % H_2O_2 by adding 30 % hydrogen peroxide (Component F) to amplification buffer (Component E) to obtain a final concentration of 0.0015 % H_2O_2. For example, add 5 μL of 30 % H_2O_2 to 995 μL of amplification buffer and then add 20 μL of this intermediate dilution (0.15 % H_2O_2) to a further 1980 μL of amplification buffer.

5.4 Summary

In this chapter, we introduced principal methods of cell staining on the micropillar/microwell chip platform, which include 3D cell-based HCI assays with fluorescent dyes as well as IF assays with primary/secondary antibodies. To demonstrate and retrofit traditional HCI assays and IF assays on the micropillar/microwell chip, the capability to optimize experimental parameters is of importance. These protocols can be easily modified to other cell staining and extended to high-throughput screening of compounds for predictive toxicology.

References

1. Charvin, G., Oikonomou, C., & Cross, F. R. (2010). Labels and probes for live cell imaging: Overview and selection guide. *Methods in Molecular Biology, 591*(2), 229–242. doi:10.1007/978-1-60761-404-3_14.
2. van Vliet, E., Danesian, M., Beilmann, M., Davies, A., Fava, E., Fleck, R., et al. (2014). Current approaches and future role of high content imaging in safety sciences and drug discovery. *ALTEX, 31*(4), 479–493. doi:10.14573/altex.1405271.
3. Buchser, W., Collins, M., Garyantes, T., Guha, R., Haney, S., Lemmon, V., et al. (2012). Assay development guidelines for image-based high content screening, high content analysis and high content imaging. In G. S. Sittampalam, N. P. Coussens, H. Nelson, M. Arkin, D. Auld, C. Austin, et al. (Eds.), *Assay guidance manual* (pp. 1–69). Bethesda, MD: Eli Lilly & Company and the National Center for Advancing Translational Sciences.
4. Zanella, F., Lorens, J. B., & Link, W. (2010). High content screening: Seeing is believing. *Trends in Biotechnology, 28*(5), 237–245. doi:10.1016/j.tibtech.2010.02.005.
5. Valeur, B., & Berberan-Santos, M. N. (2012). Autofluorescence and fluorescence labeling in biology and medicine. *Molecular Fluorescence: Principles and Applications, 479–505*.
6. Demchenko, A. P. (2008). *Introduction to fluorescence sensing*. Berlin: Springer. doi:10.1007/978-1-4020-9003-5.
7. Virus, E. D., Sobolevsky, T. G., & Rodchenkov, G. M. (2008). Introduction to fluorescent techniques. *Journal of Mass Spectrometry, 43*(7), 949–957. doi:10.1002/jms.1447.
8. Towne, D. L., Nicholl, E. E., Comess, K. M., Galasinski, S. C., Hajduk, P. J., & Abraham, V. C. (2012). Development of a high-content screening assay panel to accelerate mechanism of action studies for oncology research. *Journal of Biomolecular Screening, 17*(8), 1005–1017. doi:10.1177/1087057112450050.
9. Martin, H. L., Adams, M., Higgins, J., Bond, J., Morrison, E. E., Bell, S. M., et al. (2014). High-content, high-throughput screening for the identification of cytotoxic compounds based on cell morphology and cell proliferation markers. *PLoS ONE, 9*(2), 1–8. doi:10.1371/journal.pone.0088338.
10. Alonso-Padilla, J., Cotillo, I., Presa, J. L., Cantizani, J., Peña, I., Bardera, A. I., et al. (2015). Automated high-content assay for compounds selectively toxic to trypanosoma cruzi in a myoblastic cell line. *PLoS Neglected Tropical Diseases, 9*, 1–17. doi:10.1371/journal.pntd.0003493.
11. Sirenko, O., Hesley, J., Rusyn, I., & Cromwell, E. F. (2014). High-content assays for hepatotoxicity using induced pluripotent stem cell-derived cells. *Assay and Drug Development Technologies, 12*(1), 43–54. doi:10.1089/adt.2013.520.
12. Håkanson, M., Cukierman, E., & Charnley, M. (2014). Miniaturized pre-clinical cancer models as research and diagnostic tools. *Advanced Drug Delivery Reviews, 69–70*, 52–66. doi:10.1016/j.addr.2013.11.010.

13. Mioulane, M., Foldes, G., Ali, N. N., Schneider, M. D., & Harding, S. E. (2012). Development of high content imaging methods for cell death detection in human pluripotent stem cell-derived cardiomyocytes. *Journal of Cardiovascular Translational Research, 5*, 593–604. doi:10.1007/s12265-012-9396-1.
14. Fujisawa, S., Romin, Y., Barlas, A., Petrovic, L. M., Turkekul, M., Fan, N., et al. (2014). Evaluation of YO-PRO-1 as an early marker of apoptosis following radiofrequency ablation of colon cancer liver metastases. *Cytotechnology, 66*(2), 259–273. doi:10.1007/s10616-013-9565-3.
15. Donato, M. T., Tolosa, L., Jiménez, N., Castell, J. V., & Gómez-Lechón, M. J. (2012). High-content imaging technology for the evaluation of drug-induced steatosis using a multiparametric cell-based assay. *Journal of Biomolecular Screening, 17*(3), 394–400. doi:10.1177/1087057111427586.
16. Lakowicz, J. R. (2009). Fluorophores. *Princ Fluoresc Spectrosc, 954*. doi:10.1002/smll.201090041.
17. Lipman, N. S., Jackson, L. R., Weis-Garcia, F., & Trudel, L. J. (2005). Monoclonal versus polyclonal antibodies: Distinguishing characteristics, applications, and information resources. *ILAR Journal, 46*(3), 258–268. doi:10.1093/ilar.46.3.258.
18. Aubin, J. E. (1979). Autofluorescence of viable cultured mammalian cells. *The Journal of Histochemistry and Cytochemistry, 27*(1), 36–43.
19. Ntziachristos, V. (2010). Going deeper than microscopy: the optical imaging frontier in biology. *Nature Methods, 7*(8), 603–614. doi:10.1038/nmeth.1483.
20. Lee, K., Choi, S., Yang, C., Wu, H.-C., & Yu, J. 2013. Autofluorescence generation and elimination: A lesson from glutaraldehyde. *Chemical Communications (Cambridge, England), 49*(29):3028–3030. doi:10.1039/c3cc40799c.
21. Shaner, N. C., Steinbach, P. A., & Tsien, R. Y. (2005). Aguide to choosing fluorescent proteins. *Nature Methods, 2*(12), 905–909. doi:10.1038/nmeth819.
22. Manning, C. F., Bundros, A. M., & Trimmer, J. S. (2012). Benefits and pitfalls of secondary antibodies: Why choosing the right secondary is of primary importance. *PLoS ONE, 7*(6). doi:10.1371/journal.pone.0038313.
23. Kaku, T., Ekem, J. K., Lindayen, C., Bailey, D. J., Van Nostrand, A. W., & Farber, E. (1983). Comparison of formalin- and acetone-fixation for immunohistochemical detection of carcinoembryonic antigen (CEA) and keratin. *American Journal of Clinical Pathology, 80*(6), 806–815.
24. Levitt, D., & King, M. (1987). Methanol fixation permits flow cytometric analysis of immunofluorescent stained intracellular antigens. *Journal of Immunological Methods, 96*(2), 233–237. doi:10.1016/0022-1759(87)90319-X.
25. Kosaka, T., Nagatsu, I., JY, W., & Hama, K. (1986). Use of high concentrations of glutaraldehyde for immunocytochemistry of transmitter-synthesizing enzymes in the central nervous system. *Neuroscience, 18*(4), 975–990. doi:10.1016/0306-4522(86)90112-0.
26. Pollice, A. A., McCoy, J. P., Shackney, S. E., Smith, C. A., Agarwal, J., Burholt, D. R., et al. (1992). Sequential paraformaldehyde and methanol fixation for simultaneous flow cytometric analysis of DNA, cell surface proteins, and intracellular proteins. *Cytometry, 13*(4), 432–444. doi:10.1002/cyto.990130414.
27. Grizzle, W. E. (2009). Models of fixation and tissue processing. *Biotechnic & Histochemistry, 84*(5), 185–193. doi:10.3109/10520290903039052.Models.
28. Miething, F., Hering, S., Hanschke, B., & Dressler, J. (2006). Effect of fixation to the degradation of nuclear and mitochondrial DNA in different tissues. *The Journal of Histochemistry and Cytochemistry, 54*(3), 371–374. doi:10.1369/jhc.5B6726.2005.
29. Jamur MC, Oliver C. (2010). Permeabilization of cell membranes. In Oliver C, & Jamur MC (Eds.), *Methods and Protocols* (Vol. 588, pp. 63–66). New York: Humana Press. doi:10.1007/978-1-59745-324-0.
30. Amidzadeh, Z., Behzad Behbahani, A., Erfani, N., Sharifzadeh, S., Ranjbaran, R., Moezi, L., et al. (2014). Assessment of different permeabilization methods of minimizing damage to the

adherent cells for detection of intracellular RNA by flow cytometry. *Avicenna Journal of Medical Biotechnology, 6*(1), 38–46.
31. Ohsaki, Y., Maeda, T., & Fujimoto, T. (2005). Fixation and permeabilization protocol is critical for the immunolabeling of lipid droplet proteins. *Histochemistry and Cell Biology, 124*(5), 445–452. doi:10.1007/s00418-005-0061-5.
32. Misra, D. P., Chaurasia, S., & Misra, R. (2016). Increased circulating Th17 Cells, Serum IL-17A, and IL-23 in Takayasu Arteritis. *Autoimmune Diseases, 2016*, 7841718. doi:10.1155/2016/7841718.
33. Kopen, G. C., Prockop, D. J., & Phinney, D. G. (1999). Marrow stromal cells migrate throughout forebrain and cerebellum, and they differentiate into astrocytes after injection into neonatal mouse brains. *Proceedings of the National Academy of Sciences of the United States of America, 96*(19), 10711–10716. doi:10.1073/pnas.96.19.10711.
34. Italiano, J. E., Richardson, J. L., Patel-Hett, S., Battinelli, E., Zaslavsky, A., Short, S., et al. (2008). Angiogenesis is regulated by a novel mechanism: Pro- and antiangiogenic proteins are organized into separate platelet α granules and differentially released. *Blood, 111*(3), 1227–1233. doi:10.1182/blood-2007-09-113837.
35. Kwong, K. F., Schuessler, R. B., Green, K. G., Laing, J. G., Beyer, E. C., Bioneau, J. P., et al. (1998). Differential expression of gap junction proteins in the canine sinus node. *Circulation Research, 82*, 604–612.
36. Bianchi, L., Shen, Z., Dennis, A. T., Priori, S. G., Napolitano, C., Ronchetti, E., et al. (1999). Cellular dysfunction of LQT5-minK mutants: Abnormalities of I(Ks), I(Kr) and trafficking in long QT syndrome. *Human Molecular Genetics, 8*(8), 1499–1507. doi:10.1093/hmg/8.8.1499.
37. Thompson, K., Trowern, A., & Fowell, A. (1998). Primary rat and mouse hepatic stellate cells express the macrophage inhibitor cytokine interleukin-10 during the course of activation in vitro. *Hepatology, 28*(6), 1518–1524. doi:10.1002/hep.510280611.

Chapter 6
3D-Cultured Cell Image Acquisition

Pranav Joshi, Kyeong-Nam Yu, Emily Serbinowski, and Moo-Yeal Lee

Contents

6.1 Introduction 125
6.2 Materials 127
6.3 Protocols 127
6.3.1 Daily Operational Procedures 128
6.3.2 Parameter Setting Procedures 135
6.3.2.1 Setting the Position of Filters and the Distance of Each Step for Autofocus 135
6.3.2.2 Setting XYZ Coordinates for Different Chips and Objective Lenses 137
6.4 Summary 139
6.5 Appendix 139
References 141

6.1 Introduction

Acquiring high-content images of 3D-cultured cells for analyzing multiple cellular events is a daunting task, requiring an automated fluorescent microscope and high-throughput image analysis software. High throughput is an important feature for high-content imaging (HCI) devices to enable rapid image acquisition. Various factors come into play when dealing with the speed of image acquisition. For example, capturing large number of cells with lower magnification or reducing sample volume/size can increase data acquisition speed. In addition, reducing exposure time with the use of high intensity light sources, optimizing fluorescence staining protocols for brighter colors, and using an objective lens with relatively high numerical aperture also significantly increases the image acquisition speed [1].

M.-Y. Lee, *Microarray Bioprinting Technology*,
DOI 10.1007/978-3-319-46805-1_6

The basic components of any imaging device consist of light sources, detectors, objective lenses, and optical filters among which light sources and detectors play critical role in determining fate of any HCI system. Light sources can comprise of lamps, lasers, and light-emitting diodes (LEDs) with each of them having their own benefits and limitations. For example, lamps provide a broad excitation source from UV to IR, but should be replaced frequently due to shorter lifetime and may need realignment for optimal excitation. Lasers, on the other hand, have longer lifetime, but may not offer optimal excitation spectra for certain target fluorophores such as blue fluorescent dyes. LEDs offer both long lifetime and optimal excitation, but are expensive. Most of currently available image acquisition systems use xenon or mercury lamps as a light source due to economic feasibility and a broad range of spectra. Detectors are another important component of HCI devices with two types of detectors commonly used. These detectors are charge-coupled devices (CCDs) and photomultiplier tubes (PMTs). CCD-based detectors are widely used in HCI devices as it offers excellent acquisition speed (100 frames per second), large dynamic ranges (>20,000:1), broad spectral sensitivity (400–900 nm and higher), and high resolution (>2000 × 2000 pixels), but with reduced sensitivity to low intensity light. On the other hand, PMT-based detectors offer high sensitivity towards low intensity light, but with compromise in throughput. Multiple PMTs are used in parallel to acquire images through various fluorescent channels [1–3].

Conventional HCI devices can be divided broadly into two categories; wide-field imaging devices and confocal imaging devices. Wide-field imaging devices are basically inverted microscopes which offer an excellent image acquisition speed. Some examples of commercially available CCD-based wide-field imaging devices are ArrayScan XTI HCS Reader manufactured by Thermo Fisher, IN Cell Analyzer 2000 by GE Healthcare, ImageXpress Micro HCS system by Molecular Devices, Operetta by Perkin Elmer, WiSCAN by IDEA Bio-Medical, and S+ Scanner by SEMCO. These commercially available wide-field imaging devices commonly use lamps and LEDs as their light source. For example, ArrayScan XTI HCS Reader uses a metal halide lamp and LED, and IN Cell Analyzer 2000 uses a metal halide lamp. Similarly, ImageXpress Micro HCS system uses a xenon lamp, and Operetta uses a xenon lamp and LED. In addition, S+ scanner uses a mercury lamp, and WiSCAN uses a mercury lamp and LED. Although different types of light sources are used by various wide-field imaging devices, the difference is very little in terms of optical performance. The difference between the various wide-field imagers available arise generally from the numerical aperture of the objectives and the quality of light path [1, 4, 5].

Confocal imaging devices use light barrier with a fixed or adjustable pinhole to eliminate light in front or behind the focus plane of an objective. Confocal devices offer high resolution and improved contrast compared to wide-field devices, but it also causes reduction in light signal and only scans single points of a specimen at any given moment. Although these issues have been overcome by using technologies like laser scanning confocal microscopy (LSCM) and Nipkow spinning disk, the image acquisition speed is compromised, and there is always a chance of photobleaching [1, 4, 6, 7]. Confocal imaging devices are expensive compared to wide-field imaging devices and most of its applications are towards imaging small intracellular structures, small cells, complex 3D structures, *etc*. IN Cell Analyzer 6000 by GE Healthcare is a commonly used confocal microscope, which uses LSCM with varying apertures. It consists of four laser lines and a LED for transmitted light and a scientific complementary metal oxide semiconductor (sCMOS)-based detector. ImageXpress ULTRA developed by Molecular devices is another confocal imager consisting of four lasers and four PMTs that can be operated serially or in parallel [1, 8, 9]. Extensive information on various HCI devices and their components can be found in Assay Guidance Manual published by Eli Lilly & Company and the National Center for Advancing Translational Sciences (http://www.ncbi.nlm.nih.gov/books/NBK53196/) [1].

While other HCI devices are highly sophisticated and require trained personnel to operate the machine, the S+ Scanner offers researchers/users the ability to acquire high quality fluorescent images in a high-throughput manner without the need of complex training. This chapter aims to provide simple procedures for operating S+ Scanner to rapidly acquire 3D-cultured cell images from the micropillar/microwell chip platform.

6.2 Materials

- Micropillar/microwell chips (25 × 75 × 2 mm from Samsung Electro-Mechanics, Co. or SEMCO, Suwon, South Korea)
- S+ MicroArrayer with six solenoid valves and ceramic tips (SEMCO, Suwon, South Korea)
- S+ Scanner with four filter sets (SEMCO, Suwon, South Korea)
- Light source (Olympus, Model: U-HGLGPS)
- Micropillar chip zig (SEMCO, Suwon, South Korea)
- Microwell chip zig part I and II (SEMCO, Suwon, South Korea)
- Metal frame for the micropillar chip (SEMCO, Suwon, South Korea)

6.3 Protocols

S+ Scanner, is a wide-field imager consisting of a mercury lamp as a light source and a CCD-based detector. It is an automated fluorescence microscope with four filter channels for detecting multicolor, blue, green, and red fluorescent dyes, individually or simultaneously. The basic components of S+ Scanner are provided in Fig. 6.1. Certain precautions are required before starting the image acquisition from the micropillar/microwell chips. For example, micropillar chips stained with fluorescent dyes should be dried properly before scanning to avoid damage to the

Fig. 6.1 Mechanical components of S+ Scanner

objective lens and stored in the dark to prevent photobleaching. In addition, when acquiring the image from microwell chips containing liquid samples it should be covered with a gas-permeable sealing membrane to prevent spilling of the samples and to minimize immediate drying of the spots.

6.3.1 Daily Operational Procedures

1. Turn on the power source (by using the blue switch on the step-up transformer).
2. Turn on the S+ Scanner using the green power switch and then press the 'Reset' button (Fig. 6.2). ***Note: Pushing the 'Reset' button resets XYZ coordinates of the chip-loading deck. This step is essential to avoid malfunctioning of the S+ Scanner. Do not skip this step!***
3. Turn on the light source and the computer. Use the black switch on the back to turn on the light source and the 'ON/OFF' button in the front to turn on the lamp (Fig. 6.3). ***Note: After the lamp is turned ON, do not turn OFF for 2 min to avoid damage to the lamp. In addition, do not turn ON the lamp within 10 min of turning OFF. The operation of switches is disabled for 5 min after the lamp is turned OFF. The lamp is connected to the light source in S+ Scanner via a fiber optic cable, which needs to be placed properly to avoid damage to the fiber optic cable.***
4. Open the scanner program by double clicking the shortcut of S+ Scanner located on the desktop (Fig. 6.4).
5. The user interface of the scanner program will be opened (Fig. 6.5).
6. Make sure that the sliding door on top of the scanner is closed (Fig. 6.2).

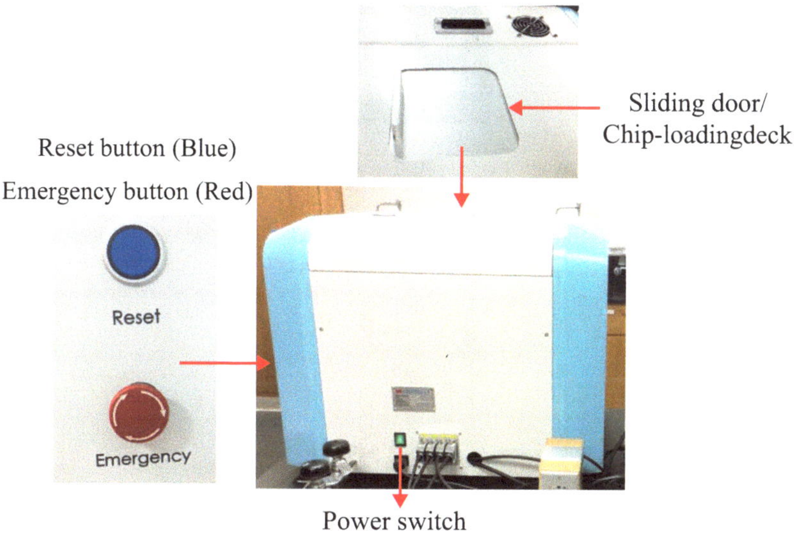

Fig. 6.2 The picture of S+ Scanner showing the power switch on the *back*, reset and emergency buttons on the *side*, and sliding door/chip-loading deck on the *top*

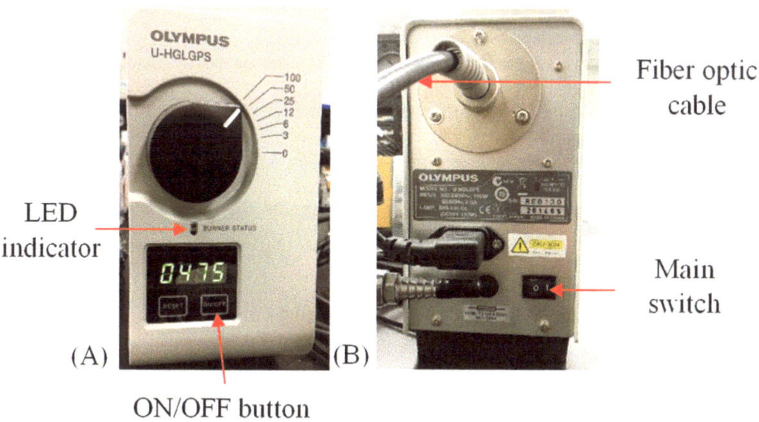

Fig. 6.3 Pictures of the light source: (**a**) The front side with ON/OFF and Reset buttons and a LED indicator displaying the burner status and (**b**) The back side with the main switch and the fiber optic cable

7. In the 'Home' window, click the 'Home' icon in the 'XYZ Move' tab to activate the 'Chip loading' and 'Live' icon (Fig. 6.6).
8. Click 'Chip loading' icon in the 'XYZ Move' tab to bring the chip-loading deck in position for loading the micropillar chip zig (Fig. 6.6).

Fig. 6.4 The shortcut of S+ Scanner program

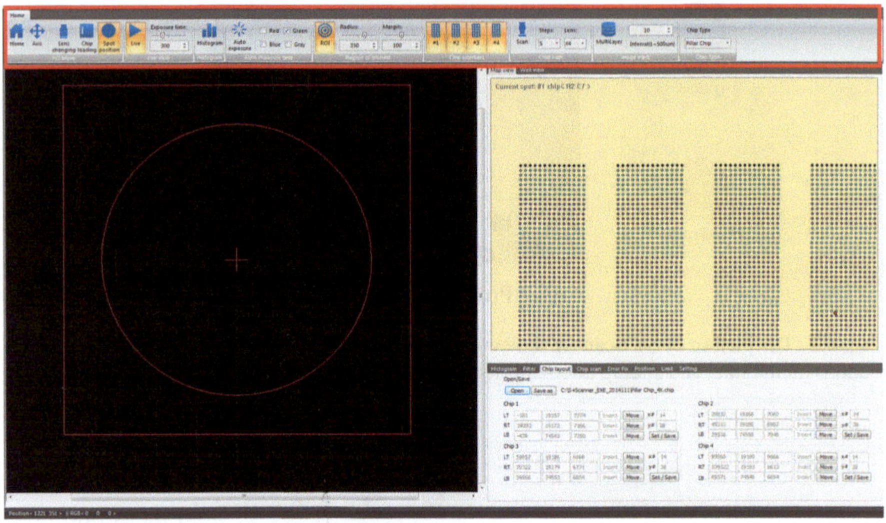

Fig. 6.5 The user interface of S+ Scanner program showing 'Home' window highlighted in a *red* box

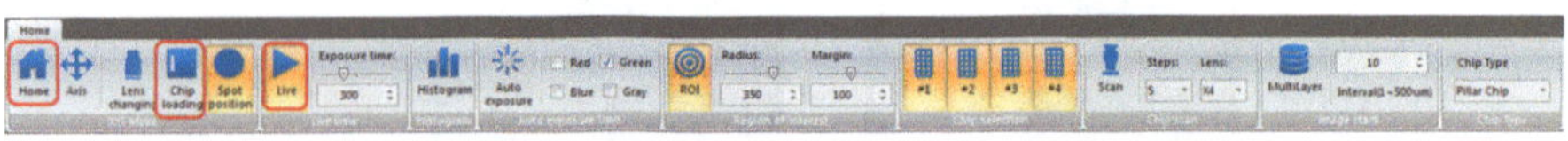

Fig. 6.6 'Home' window in the user interface of the scanner program

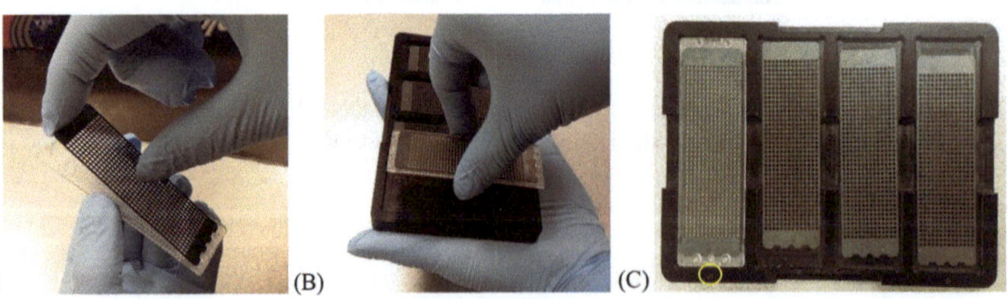
(A) (B) (C)

Fig. 6.7 Preparation of micropillar chips before loading on the chip-loading deck for image acquisition. (**a**) Placing a metal frame on the micropillar chip at an angle of 45°, (**b**) Inserting all the micropillars into the holes on the metal frame, (**c**) Placing the micropillar chip with the metal frame on the magnetic micropillar chip zig. The *circle* highlighted in *yellow* indicates the bottom of the zig aligned with the bottom of the micropillar chip.

9. In order to place micropillar chips on the micropillar chip zig and load it into the chip-loading deck, follow the steps below (Fig. 6.7).
   (a) Align a metal frame on a micropillar chip by holding the metal frame at an angle of 45° against the edge of the micropillar chip (Fig. 6.7A). The dents at the bottom of the metal frame should be perfectly aligned with the circles at the bottom of the micropillar chip.
   (b) Slowly drop the metal frame so that all the micropillars go into the holes on the metal frame (Fig. 6.7B).
   (c) Gently insert the micropillar chip with the metal frame into the magnetic micropillar chip zig by holding the side of the chip with two fingers (Fig. 6.7B).
   (d) The micropillar chip will be laid flat and held firmly while scanning due to strong magnets in the micropillar chip zig (Fig. 6.7C). ***Note: The strong magnetic attraction between the zig and the metal frame may impose severe force on the micropillar chip. Therefore, make sure that all the micropillars go into the holes on the metal frame properly before placing it on the zig.***
10. Open the sliding door and load the micropillar zig with the chips into the chip-loading deck by inverting the zig (Fig. 6.8). ***Note: The symbol (4) on the zig faces the same direction as the symbol on the chip-loading deck.***
11. In case of scanning microwell chips, the chip should be sealed with a gas-permeable membrane to prevent water evaporation in the microwells and spilling of the sample over the objective lens.
12. A microwell chip zig consists of two parts—part I is an open metal frame for placing microwell chips and part II is a magnetic frame that can hold the microwell chips firmly while scanning due to strong magnets (Fig. 6.9).

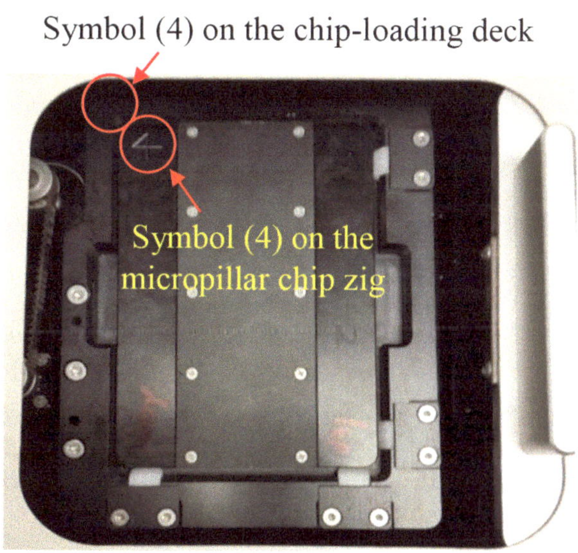

Fig. 6.8 The micropillar chip zig with micropillar chips loaded into the chip-loading deck. The symbol (4) on the micropillar chip zig should be aligned in the same direction with the symbol (4) on the chip-loading deck.

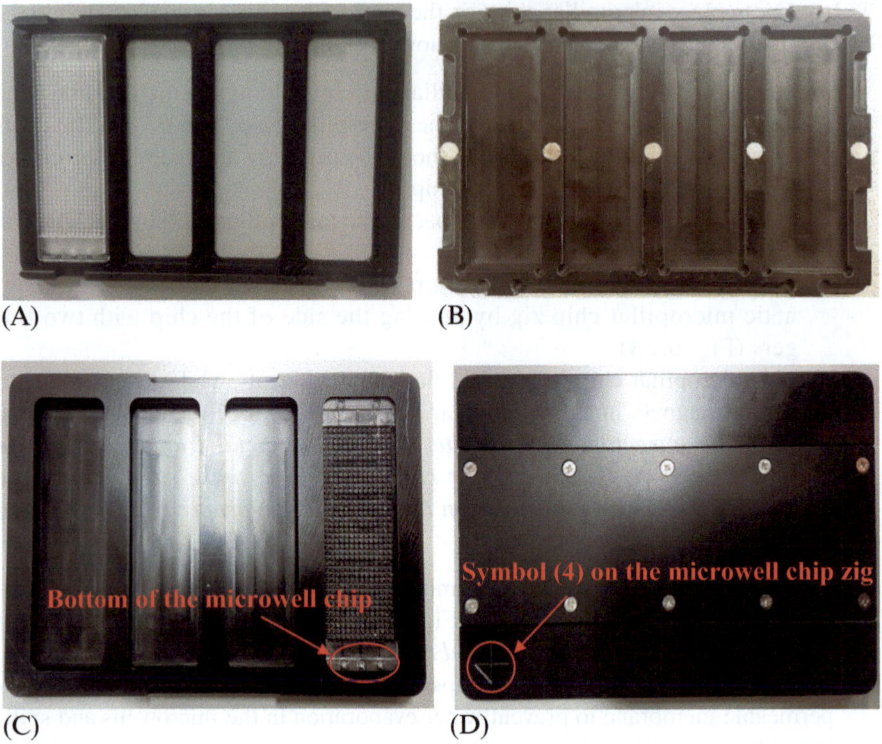

Fig. 6.9 (**a**) A microwell chip inserted in the part I microwell chip zig, (**b**) The front side of the part II microwell chip zig, (**c**) The part I zig combined with the part II zig, and (**d**) The back side of the part II zig combined with the part I zig. The symbol (4) on the back of the part II zig highlighted in red should be aligned with the bottom of the microwell chips inserted.

13. In order to place microwell chips on the microwell chip zig and load it into the chip-loading deck, follow the steps below (Fig. 6.9).
    (a) Place and insert microwell chips in the part I microwell chip zig (Fig. 6.9A) so that the back side of the microwell chip is exposed for scanning while the front side of the chip is sealed with the gas-permeable membrane, facing the part II microwell chip zig (Fig. 6.9B).
    (b) Take the part II zig and place it on the part I zig so that the symbol [4] on the part II zig is aligned with the bottom of the microwell chips (Fig. 6.9C, D). ***Note: The microwell chip will be laid flat and held firmly while scanning due to strong magnets in the microwell chip zig.***
14. Click 'Open' in the 'Chip layout' tool bar (Fig. 6.10). A window displaying a list of chip files will appear (Fig. 6.11). ***Note: These chip files contain XYZ coordinates for the micropillar chip and the microwell chip at different objective lenses.***

Fig. 6.10 The display of the 'Chip layout' tool bar

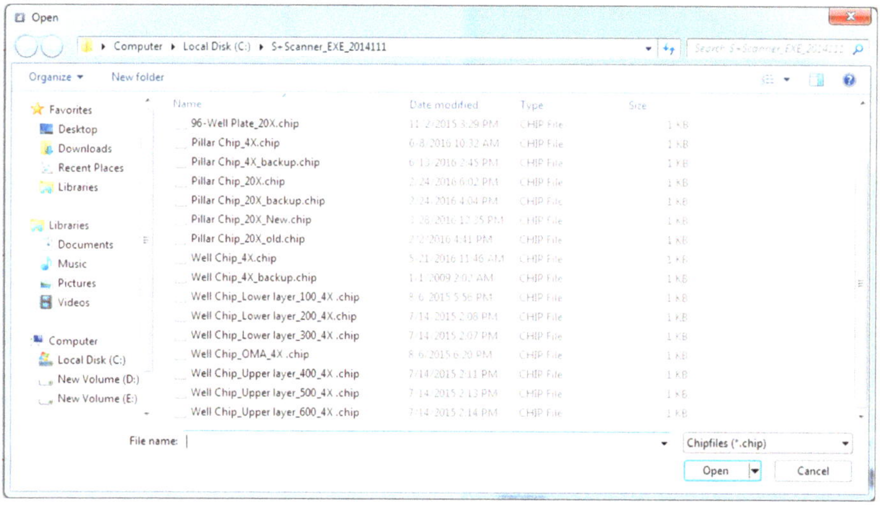

Fig. 6.11 Window displaying a list of chip files

15. Select a desired chip file from the list by clicking 'Open' in the 'Chip layout' window (Fig. 6.11). ***Note: Refer to parameter setting procedures to set and change XYZ coordinates in a given chip file.***
16. In the 'Home' window, click the 'Spot position' icon in the 'XYZ Move' tab to enable accurate selection of individual spots (i.e., micropillars or microwells) for live view (Fig. 6.12).
17. Select the number of chips being scanned from the 'Chip selection' tab in the 'Home' window (Fig. 6.12).
18. In the 'Live view' tab, click the 'Live' icon to observe a real-time cell image and check an optimum exposure time by dragging the cursor left to right in the 'Exposure time' section or manually entering the value in the 'Exposure time' section (Fig. 6.12).

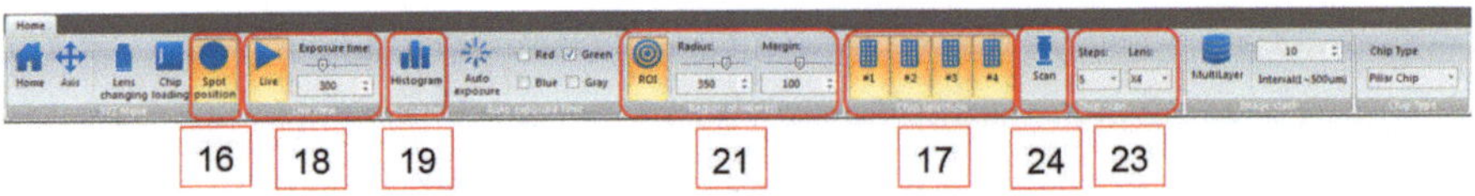

Fig. 6.12 'Home' window of S+ Scanner program

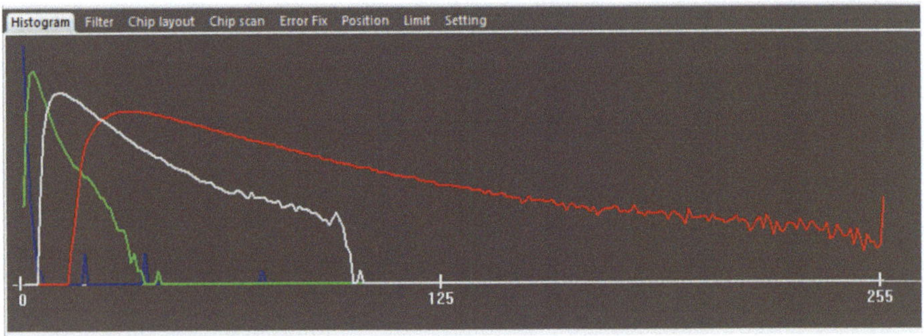

Fig. 6.13 'Histogram' window displaying the histogram of the *red*-fluorescent image from a *red* filter. The *red* intensity is higher than 255, meaning that the exposure time is a bit too high

19. Click the 'Histogram' icon in the 'Home' window to enable the display of fluorescence histogram (Fig. 6.12).
20. To find a proper exposure time, use the 'Histogram' tool bar to analyze the wavelength and also visually inspect the real-time cell image. An exposure time which gives a histogram between 80 and 90 % of maximum pixel range (i.e., 255) for the given channel is considered as an optimum exposure time (Fig. 6.13). ***Note: Do not select too high exposure time as photobleaching of fluorescence might occur at higher exposure time.***
21. In the 'Region of interest' tab, click 'ROI' icon to display the region of interest (ROI) in the image window and use the 'Radius' and 'Margin' cursor or enter the value to increase and decrease the size of the ROI (Fig. 6.12).
22. In the 'Chip scan' tool bar, click 'Set' to activate the setting window for each chip. Assign a label for each chip in the 'Folder' section, set the exposure time in 'Exposure' section, and select the desired filter from the drop-down list in the 'Filter' section. (Fig. 6.14). ***Note: A single chip can be scanned using four different filters and exposure setting depending on the user requirement. Make sure that the rest of the drop-down list in 'Filter' section is set to 'UNUSED' when scanning the chip with less than four filters setting. Select 'Multiband' filter for acquiring images from cells stained with multiple fluorescent dyes. The drop-down list in 'Auto-Exp' field is usually set to 'UNUSED' when scanning the chip with a user-defined exposure setting.***

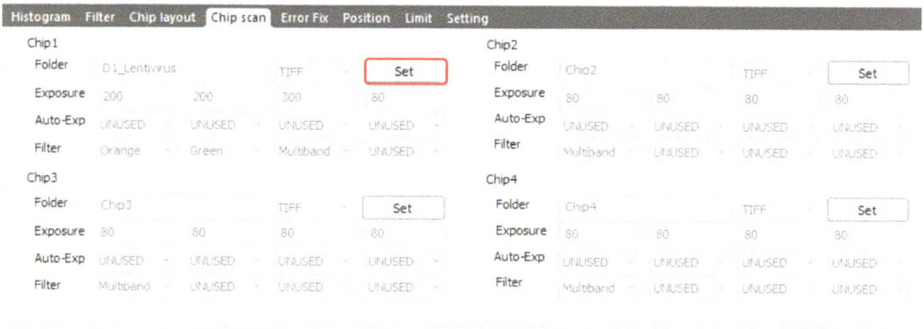

Fig. 6.14 The display of the 'Chip scan' tool bar

23. Select the magnification of an objective lens used for image acquisition from the drop down list in the 'Lens' field and select the number of steps for autofocus from the drop down list in the 'Steps' field in the 'Chip scan' tab (Fig. 6.12). ***Note: Select a larger number of steps when the chip is bent or not placed flat. It will cover a larger distance in the Z direction to obtain an optimum image in focus.***
24. Click the 'Scan' icon to begin chip scanning.
25. The images obtained from individual filters are saved in multiple folders with the name of the specific filter and the exposure time used for that filter in the format of 'filter_exposure time' (e.g., Multiband_150). These folders are stored inside another folder with the name of the chip provided by users (Refer to Step 22). The user-named folders for multiple chips are stored in a big folder which is labeled in the format of 'year-month-day' (e.g., 2016-06-01).
26. After the scanning is done, remove the chips from the scanner and store the scanned chips in the dark until disposed.
27. Close the scanner software and turn off the computer, the S+ Scanner, and the light source (Fig. 6.3). ***Note: To switch off the light source, first press the 'ON/OFF' button in the front until the blue LED turns off and the countdown begins from 300 on the digital display. Wait until the countdown ends and then turn off the main switch of the light source located on the back.***

6.3.2 *Parameter Setting Procedures*

6.3.2.1 Setting the Position of Filters and the Distance of Each Step for Autofocus

This tool bar displays all the list of filters available for 'Live' view of cells and chip scanning, allows to set and change the position of the filters, and also allows to modify the distance of each step for autofocus (Refer to Step 22). ***Note: The number of steps and the distance of each step are important to obtain an optimum image in focus using the autofocus function. The S+ Scanner takes multiple***

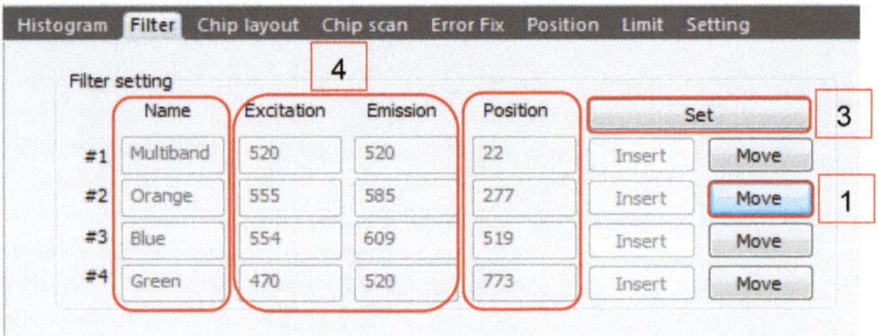

Fig. 6.15 The display of the 'Filter' tool bar

pictures in the Z direction but save only one best-looking image. Ideally, the distance of each step should be equal to the height of cell spheroids.

1. Click 'Move' button to select the desired filter, which has predetermined excitation and emission spectra for image acquisition (Fig. 6.15 and Appendix). For example, clicking the 'Move' button for the 'Orange' filter enables 'Live' view of cells/spheroids stained with orange/red fluorescent dyes in the excitation/emission range of the 'Orange' filter.
2. Filters are set to the position displayed on the 'Position' column by default. Change or reset the position of individual filters only when the filter position is misplaced due to machine malfunctioning or user errors.
3. To change or reset the filter position, click the 'Set' button to activate the 'Filter setting' window and enter the position of a respective filter so that 'Live' cell image is visible uniformly without distortion (Fig. 6.15). ***Note: The filter position has to be increased or decreased by few units only if needed. For example, the 'Multiband' filter position has to be increased or decreased to 23 or 21, respectively.***
4. The 'Excitation' and 'Emission' columns are used to calculate and set the distance of each step for autofocus (Fig. 6.15). ***Note: It doesn't indicate actual excitation and emission wavelengths.*** When we select the number of 'Steps' in the 'Chip scan' tab for autofocus (Fig. 6.12), it uses the distance of each step (i.e., the thickness of one step) calculated from the following formula.

$$\text{Distance of each step}\,(\mu m = \left[\text{Emission wavelength} / \left(\text{Lens}\,NA\right)^2 \times 1000)\right] \times \left(2/3\right)$$

where NA is numerical aperture.

For example, when 4× and 20× lenses are used at the emission wavelength of 500 nm, the distance of each step is calculated as follows:

$$\text{Step distance for } 4 \times \text{lens} = \left[500 / \left(0.13 \times 0.13 \times 1000\right)\right] \times \left(2/3\right) = 19.7\,\mu m$$

$$\text{Step distance for } 20\times \text{lens} = \left[500/(0.5\times 0.5\times 1000)\right]\times(2/3) = 1.3\,\mu m$$

where the numerical apertures for the 4× lens and the 20× lens are 0.13 and 0.5, respectively.

Note: The recommended distance of each step is 10–20 μm (e.g., 400 nm emission wavelength with 4× lens) for 3D-cultured cells and 2–5 μm (e.g., 120 nm emission wavelength with 4× lens) for 2D cell monolayers. For a small step distance, it will require a large number of 'Steps' for better autofocus. The emission wavelength setting can be changed for each filter used and the morphology of cells.

6.3.2.2 Setting XYZ Coordinates for Different Chips and Objective Lenses

The 'Chip layout' tool bar allows to design the scanning layout (i.e., the number of micropillars or microwells per chip) and determine specific XYZ coordinates of left top (LT), right top (RT), and left bottom (LB) corners of the chip for accurate image acquisition.

1. Click 'Open' in the 'Chip layout' tool bar to select a desired chip file from the list (Fig. 6.16).
2. Select a desired chip file and click 'Open' in the window. For example, the layout of the micropillar chip at 4× magnification lens is selected (Fig. 6.17).
3. To change the X, Y, and Z coordinates of the chips in the desired chip file, click 'Set/Save' button among four chips loaded (i.e., Chip 1, Chip 2, Chip 3, and Chip 4) (Fig. 6.16). ***Note: This will allow one to change the X, Y, and Z coordinates of LT, RT, and LB corners of the chip selected.***
4. To read accurate XYZ coordinates, click the 'Axis' icon in the 'XYZ Move' tab in 'Home' window to open the 'Axis' window (Fig. 6.18).
5. In the 'Live view' tab, click the 'Live' icon to observe a real-time cell image. ***Note: We need to place three micropillars or microwells at the LT, RT, and LB corners in the center of the ROI.***

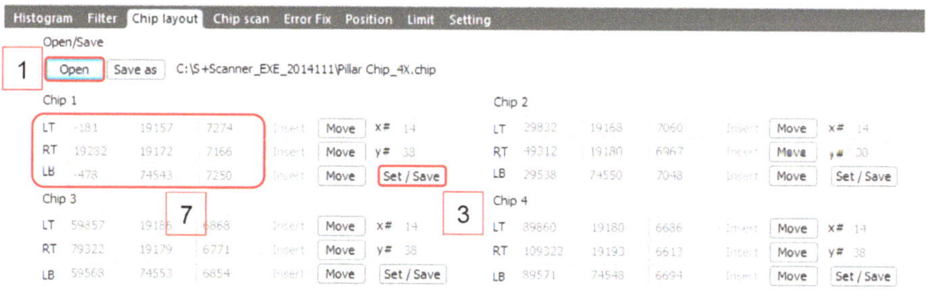

Fig. 6.16 The display of the 'Chip layout' tool bar

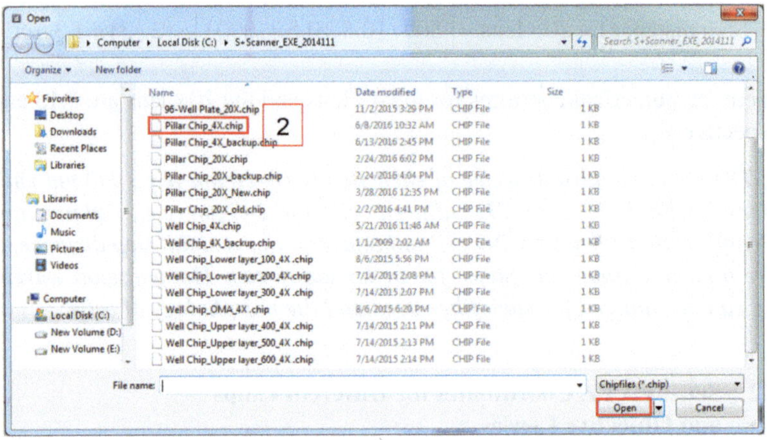

Fig. 6.17 Window displaying the list of chip files

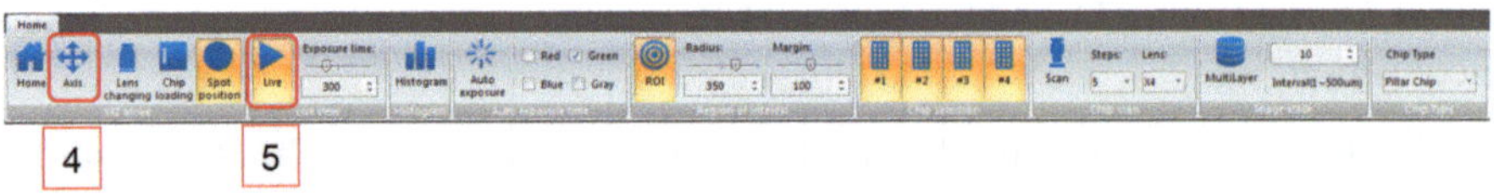

Fig. 6.18 'Home' window showing the 'Axis' icon highlighted in *red* in the 'XYZ Move' tab

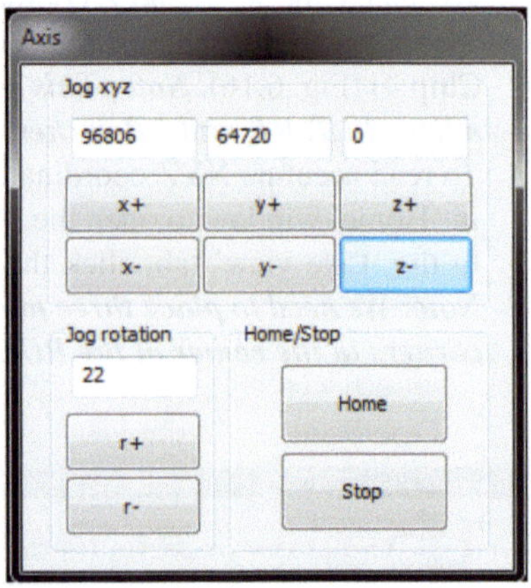

Fig. 6.19 The display of 'Axis' window

6. In the 'Axis' window (Fig. 6.19), increase/decrease X, Y, and Z coordinates by clicking the x+/x−, y+/y−, z+/z− buttons respectively to identify three micropillars or microwells at the LT, RT and LB corners of the chip. ***Note: The changes in the XYZ coordinates will be reflected immediately in the live image of cells.***
7. Select the optimum XYZ coordinates at which the image is properly focused, and enter those coordinates in the 'Chip layout' tool bar (Fig. 6.16).

6.4 Summary

Imaging technology is the key determinant for a successful HCI assay. An ideal HCI system enables the user to acquire high-resolution images of fluorescently labeled cells at a high speed in real time. In this chapter, we introduced some commonly available HCI devices and provided simple procedures for acquiring high-content images in a high-throughput fashion from the micropillar/microwell chip platform using the S+ Scanner. These procedures can be further modified to acquire high-content images from traditional HTS platforms such as 96- and 384-well plates.

6.5 Appendix

In the S+ Scanner, the following four filter sets are installed.

1. Multiband filter

Model (Semrock)	DA/FI/TR/Cy5-A-000 (Multi-band)

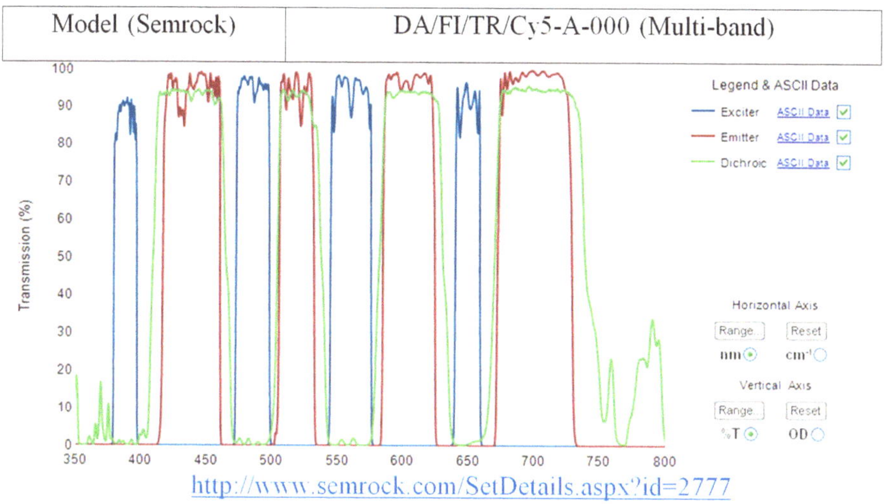

http://www.semrock.com/SetDetails.aspx?id=2777

2. Orange filter

Model (Semrock)	TxRed-4040C-000 (RED)

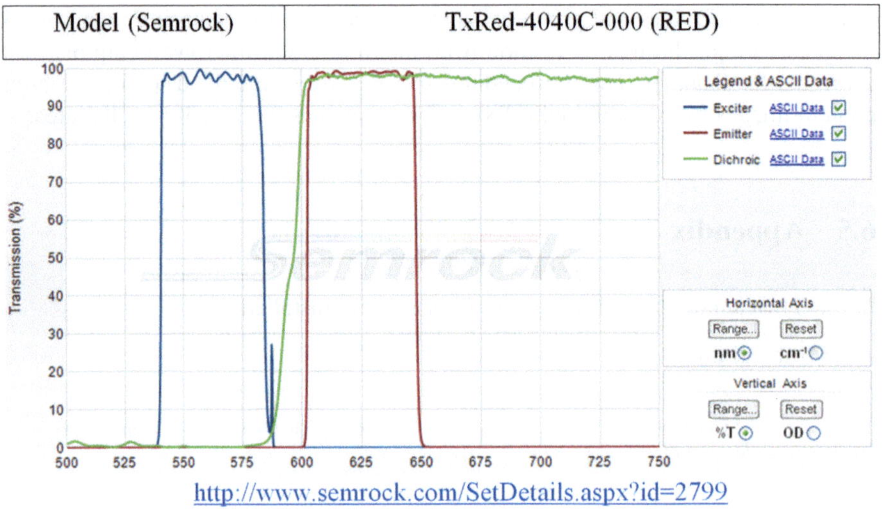

http://www.semrock.com/SetDetails.aspx?id=2799

3. Blue filter

Model (Semrock)	DAPI-5060C-000 (Blue)

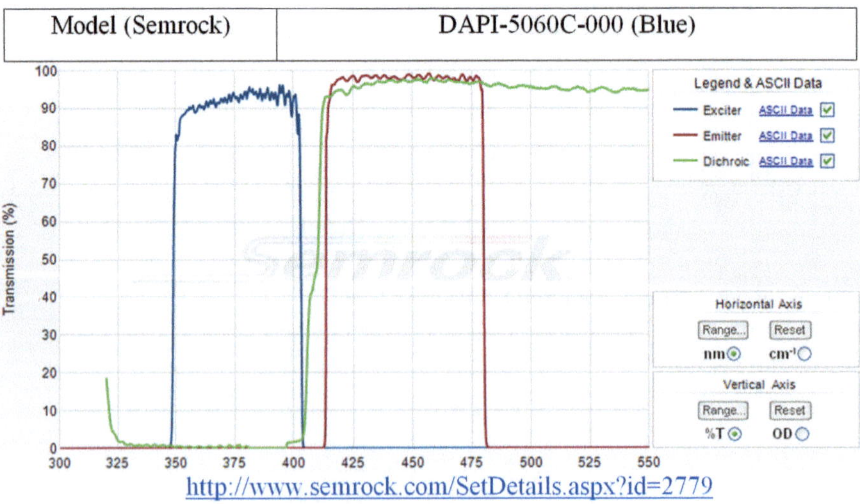

http://www.semrock.com/SetDetails.aspx?id=2779

4. Green filter

Model (Omega)	XF404 (Green)

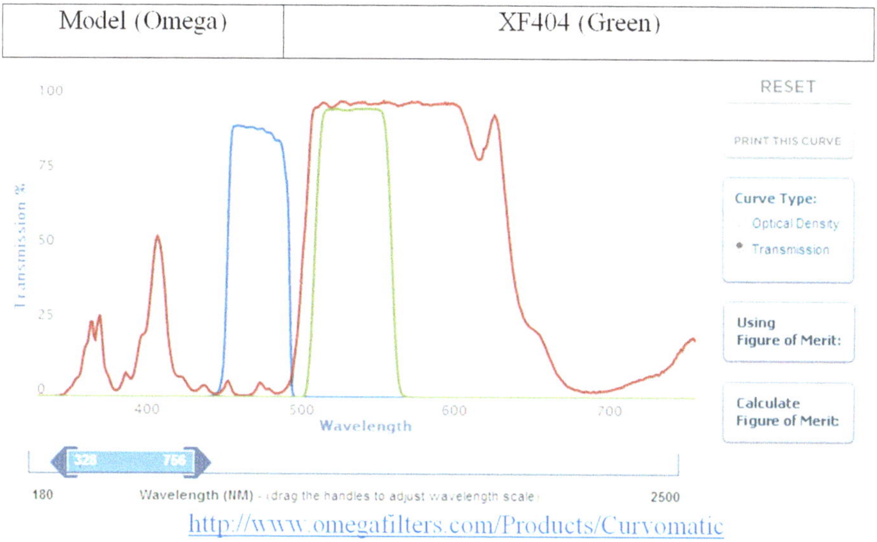

http://www.omegafilters.com/Products/Curvomatic

References

1. Buchser, W., Collins, M., Garyantes, T., Guha, R., Haney, S., Lemmon, V., et al. (2012). Assay development guidelines for image-based high content screening, high content analysis and high content imaging. In S. GS, C. NP, H. Nelson, M. Arkin, D. Auld, C. Austin, et al. (Eds.), *Assay guidance manual.* (pp. 1–69). Bethesda, MD: Eli Lilly & Company and the National Center for Advancing Translational Sciences.
2. Talbot, C. B., McGinty, J., Grant, D. M., McGhee, E. J., Owen, D. M., Zhang, W., et al. (2008). High speed unsupervised fluorescence lifetime imaging confocal multiwell plate reader for high content analysis. *Journal of Biophotonics, 1*(6), 514–521. doi:10.1002/jbio.200810054.
3. Celli, J. P., Rizvi, I., Blanden, A. R., Massodi, I., Gidden, M. D., Pogue, B. W., et al. (2014). An imaging-based platform for high-content, quantitative evaluation of therapeutic response in 3D tumour models. *Scientific Reports*. doi:10.1038/srep03751.
4. Romano, S. N., & Gorelick, D. A. (2014). Semi-automated imaging of tissue-specific fluorescence in Zebrafish Embryos. *Journal of Visualized Experiments, 87*, e51533. doi:10.3791/51533.
5. Bickle, M. (2010). The beautiful cell: High-content screening in drug discovery. *Analytical and Bioanalytical Chemistry, 398*, 219–226. doi:10.1007/s00216-010-3788-3.

6. Jahr, W., Schmid, B., Schmied, C., Fahrbach, F. O., & Huisken, J. (2015). Hyperspectral light sheet microscopy. *Nature Communications, 6*, 1–7. doi:10.1038/ncomms8990.
7. Reynaud, E. G., Peychl, J., Huisken, J., & Tomancak, P. (2015). Guide to light-sheet microscopy for adventurous biologists. *Nature Methods, 12*(1), 30–34. doi:10.1038/nmeth.3222.
8. Rae Chi, K. (2012). High on high content: A guide to some new and improved high-content screening systems. *The Scientist Magazine*, 1–7.
9. Starkuviene, V., & Pepperkok, R. (2007). The potential of high-content high-throughput microscopy in drug discovery. *British Journal of Pharmacology, 152*(1), 62–71. doi:10.1038/sj.bjp.0707346.

Chapter 7
High-Content Image Analysis

Sean Yu, Pranav Joshi, Dong Woo Lee, and Moo-Yeal Lee

Contents

7.1 Introduction 143
7.2 Materials 145
7.2.1 ImageJ 145
7.2.2 S+ Chip Analysis 145
7.3 Protocols 146
7.3.1 3D Cell Image Analysis with ImageJ 146
7.3.2 Examples of Image Processing with ImageJ 149
7.3.2.1 Hue Filter 149
7.3.2.2 Background Subtraction, Brightness Filter, and Region of Interest (ROI 150
7.3.2.3 Outlier Exclusion 151
7.3.2.4 The Performance of the Plugin 151
7.3.3 Image Deconvolution 152
7.3.3.1 Performance of Deconvolution 154
7.3.4 Plotting Dose Response Curves with S+ Chip Analysis 154
7.4 Summary 159
References 159

7.1 Introduction

High-content imaging (HCI) and image processing of cells grown in 3D pose a significant challenge because 3D cells are not grown in a single focal plane, and the cell culture systems are often incompatible with traditional microscopes. Confocal microscopy is an important tool for imaging 3D cells due to its ability to acquire high definition images at various optical sections. However, the low scanning rate and depth induce low throughput in image acquisition and also incur additional problems such as photobleaching or phototoxicity [1–3]. To alleviate these issues, miniaturized 3D cell cultures on a micropillar/microwell chip platform have been

M.-Y. Lee, *Microarray Bioprinting Technology*,
DOI 10.1007/978-3-319-46805-1_7

demonstrated, which facilitate high-throughput spheroid cultures in hydrogels while offering better imaging capabilities. The whole sample depth can fit within the focus depth of a normal objective lens due to the small dimensions.

For more accurate image acquisition, one needs to take precautions of carrying out experiments, which will minimize defects, accentuate features, and help in extracting meaningful information from the cell images acquired. For example, it is absolutely important to avoid dust particles while rinsing and staining cells, which can be a source of artifacts. In addition, it is necessary to avoid excitation/emission spectrum overlapping among different fluorescent probes to obtain clear separation of fluorescence signals when multiple dyes are used simultaneously. When acquiring images, various parameters of the imaging system such as exposure time and focus positions should be optimized to avoid photo saturation and blurry image. After acquiring images, the images can be corrected to account for various imperfections, including high background noise, dust, imaging artifacts, uneven illumination, and blurring. This is followed by segmentation to extract a particular feature in the image and quantify the information. For example, to calculate the average size of cells, we need to identify what constitutes a cell in the images—this may involve understanding the morphology of a cell, the stained color of the cell, the size range of cells, *etc*. Once these features have been properly identified and segmented, we can quantify the information, transport and warehouse the data, and then manipulate and process the data.

Various biological image analysis tools such as CellProfiler, Fiji, BioImageXD, EBImage, CellCognition, and other commercial software sold with HCI instruments are available for high-content image analysis. These tools measure the features and extract the quantitative values for analysis. If the experiment requires techniques that are highly instrumentation-specific, the complementary software can serve as an excellent starting point. Features measured typically consist of cell count, size and shape, intensity, texture, location, and clustering. CellProfiler for instance was developed for measuring cellular features such as the number of cells, size, and other morphological features, protein level per cell, and cell cycle distribution. Major steps of image processing and data analysis include image processing, object segmentation, feature extraction and selection, and statistical analysis.

An example of image acquisition and analysis tools for 3D cells is the processing and analysis software that comes with the S+ Scanner, including S+ Chip Analysis. Other tools which can be highly subject-specific or research group-specific can be found in research articles. There are groups and organizations that maintain and curate these tools within their specialties, and the listings can often be found online. For example, the Neuroimaging Tools and Resource Clearinghouse hosts and categorizes various tools pertaining to neuroimaging at nitrc.org. General image analysis tools can also be modified and extended to address a broader host of problems. Most notably, ImageJ is one of such tools that has been widely adopted due to open source and extensibility. There are countless plugins developed for ImageJ. In addition, there are novel image analysis platforms that base their architecture on ImageJ such as Icy and Vaa3D. Due to its incredible versatility and widespread use, ImageJ with an in-house plugin will be discussed extensively in the following section. The landscape of imaging processing and analysis is broad and complicated; please refer to Biological Imaging Software Tools by Kevin W Eliceiri *et al.* for an excellent overview [4]. In this chapter, we summarize protocols for analyzing images of 3D-cultured cells on the micropillar/microwell chip obtained by staining with multiple fluorescent dyes.

7.2 Materials

ImageJ is used to extract fluorescent intensity from 3D-cultured cell images, and S+ Chip Analysis is used to plot dose response curves from the extracted fluorescence data in high throughput.

7.2.1 ImageJ

Among several image analysis software available, including Image-Pro, CellProfiler, and the Image Processing Toolbox for MATLAB, we determined to use ImageJ with an in-house BioPrinting plugin. ImageJ is a Java-based image analysis tool that is designed for biological research and is free, open-source, lightweight, and extensible. Its extensibility is an enormous advantage as it allows researchers to develop macros or plugins specific to their research applications. We downloaded version 1.49v ImageJ (https://imagej.nih.gov/ij/download.html) and generated an in-house plugin for high-throughput 3D cell image analysis by modifying existing plugins and ImageJ-native functionality (http://rsb.info.nih.gov/ij/plugins/). The plugin can apply a hue filter, a brightness filter, subtract background, select a region of interest, ignore problematic outliers, save the processed images, and save the quantified fluorescence data. After processing cell images obtained from the micropillar/microwell chip using the BioPrinting plugin, the resulting fluorescence data are stored in an analysis-friendly format, which are further analyzed with S+ Chip Analysis for plotting dose response curves and calculating IC_{50} values, as well as Minitab or R for statistical analysis. The IC_{50} value represents the concentration of a compound at which cell growth is inhibited by 50 %.

7.2.2 S+ Chip Analysis

S+ Chip Analysis is an image processing and data analysis software particularly developed for visualizing and analyzing images from microarray chip platforms. This software is capable of automatic detection of spots on the microarray chip, and removal of outlier spots. It can analyze multiple images in a single batch and generate dose response curves in multiple formats along with curve fitting. In addition, this software exports the results in MS-Word format for easy access and storage of data. All the above features make it an ideal software for high-throughput compound toxicity analysis.

7.3 Protocols

7.3.1 3D Cell Image Analysis with ImageJ

1. Download ImageJ from https://imagej.nih.gov/ij/download.html and install it in your computer.
2. Place the BioPrinting plugin in the plugin folder within the ImageJ installation folder to prevent the user from having to relocate the plugin every time they want to run it.
3. Run ImageJ, locate the 'BioPrinting' tab under the 'Plugins' menu, and select the 'BioPrinting' plugin (Fig. 7.1).
4. The Plugin will ask the user to select a source directory—select the directory where the scanned images are stored (Fig. 7.2). ***Note: Multiple images in the folder are processed simultaneously in a single batch.***
5. In the 'BioPrinting Image Analysis' window, configure the processing variables (Fig. 7.3).
6. Provide a name for a set of raw images in the 'Label' field. This name will be used for the data files containing the fluorescence information as well as the folder containing that data files.
7. If cell images contain multiple distinct dyes, the plugin can separate them using a hue filter - specify the hue range by designating a minimum and maximum hue value for one of the dyes and then re-run the plugin with a different hue range for

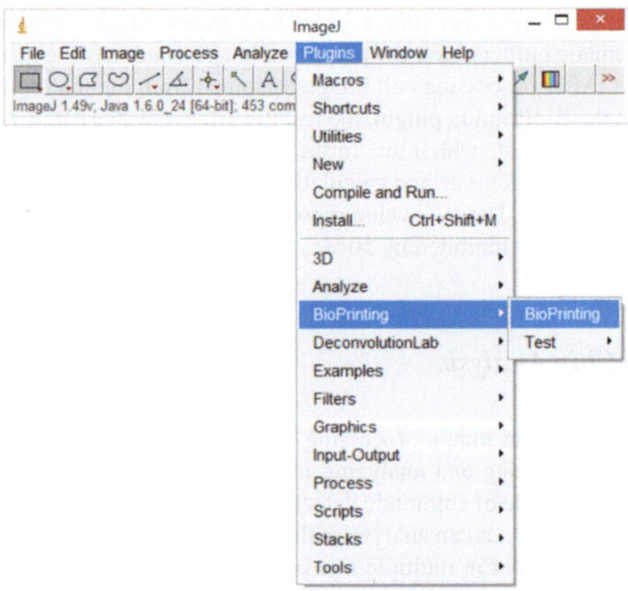

Fig. 7.1. Selection of the BioPrinting plugin from the Plugins menu

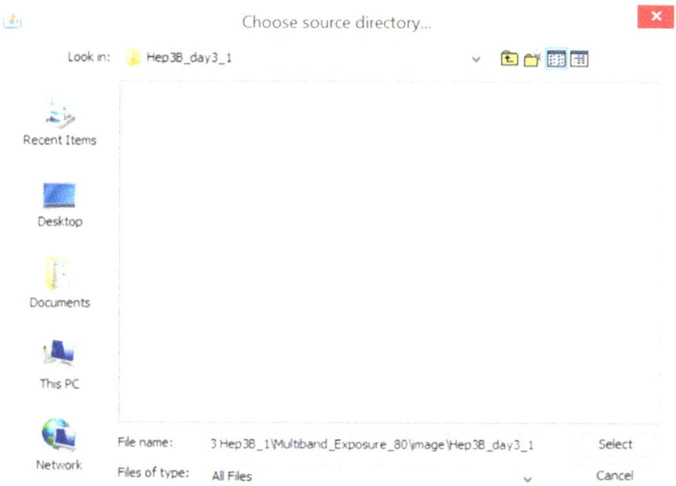

Fig. 7.2. Selecting a directory with raw images to process

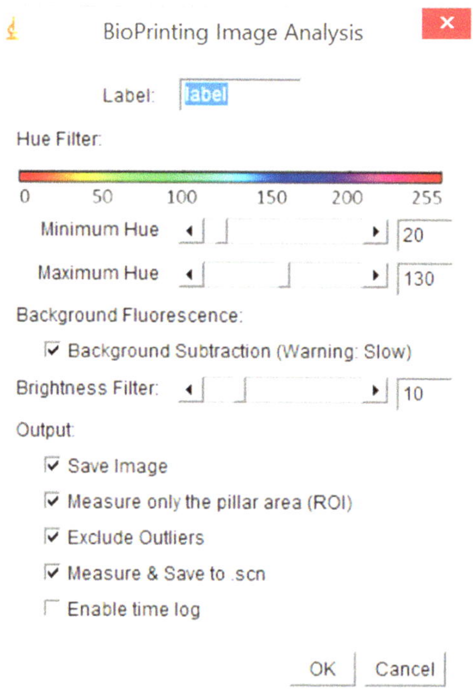

Fig. 7.3. BioPrinting plugin interface

the other dye to get separate quantitative data for two separate dyes in the same image set. As for the hue range, calcein AM, for example, has a working range of 20–130 (Table 7.1).

Table 7.1. Hue ranges of fluorescent dyes with *green*, *red*, and *blue* colors.

Dyes	Fluorescent color	Minimum hue	Maximum hue
Calcein AM	Green	20	130
Ethidium homodimer-1	Red	0	50
Hoechst 33342	Blue	120	200
Tetramethyl rhodamine methylester (TMRM)	Red	0	50

8. Reduce the background by applying a uniform brightness filter in a 0–50 scale. ***Note: If the images exhibit very low or no background fluorescence, then no brightness filter is required (brightness filter value 0). If there is significant background fluorescence, a brightness filter value of up to 50 is recommended. Be wary that fluorescence from cells will be lost if the brightness filter value is too high compared to the fluorescence signal of the stained cells.***
9. If the background fluorescence is significant even after the application of the brightness filter, and especially if the images are beset with high frequency background noise, check the 'Background Subtraction' checkbox to apply the rolling ball background subtraction method [5]. ***Note: This background subtraction method is slow, so use it only when needed.***
10. Save the processed images using the 'Save Image' checkbox, or forgo that option to reduce plugin runtime.
11. Check the 'Measure only the pillar area (ROI)' checkbox to analyze a particular region of interest. ***Note: As with most culture platforms, if the region of interest is circular, this option can allow you to only measure fluorescence from the circular central area.***
12. Select 'Exclude Outliers' to ignore the data points where the fluorescence value is two or more standard deviations away from the mean (Fig. 7.9). ***Note: Due to dust or optical aberrations, there may be huge spikes of fluorescence. In addition, certain procedural failures can lead to spot detachment from micropillars thereby resulting in lack of cells.***
13. Save quantified fluorescence data. Two files are generated (Figs. 7.4). The first one includes fluorescence (i.e., raw integrated fluorescent intensity from image processing) in a 14 by 38 matrix to match the layout of the micropillar/microwell chip. Although this format is intuitive because it matches the layout of the chip, it requires manipulation before it can be used in other statistical packages. For that purpose, another file stores the data in a 2 column format, where the first column indicates the block and the second stores the fluorescence. ***Note: The chip is divided into 6 blocks, each block with a 14 by 6 matrix, which is a common experimental layout on the chip for testing compound toxicity (e.g., 6 dosages with 14 replicates).***
14. The time log option is available for those who want to tinker with the plugin and improve its performance. It will log how long it takes to perform each processing step and how long it takes to process each image.

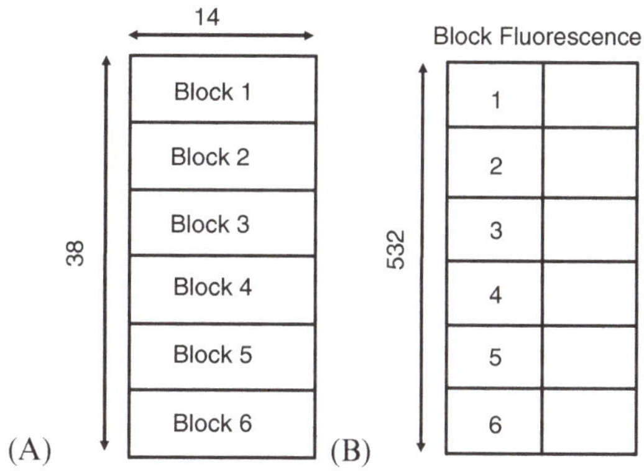

Fig. 7.4. (**a**) The *.scn data output file that matches the layout of the micropillar/microwell chip (14 × 38 data array) and (**b**) The *.txt output is in 6 block format (1 × 532 data array) for ease of use elsewhere such as statistical software.

7.3.2 Examples of Image Processing with ImageJ

Note: The following images were selected for their undesirable features to demonstrate the performance of the BioPrinting plugin and are not representative of images acquired from the micropillar/microwell chip platform using the S+ Chip Scanner.

7.3.2.1 Hue Filter

As in live/dead assays with calcein AM and ethidium homodimer-1, fluorescent images obtained from a multiband filter can have multiple peaks of fluorescence, but one may want to acquire information about one type of fluorescence at a time. In addition, optical aberrations and debris are often present in hues distinct from the emission spectra of the stain. In these cases, we can apply a hue-based band-filter through the ImageJ BioPrinting plugin to isolate the desired fluorescence (Fig. 7.5).

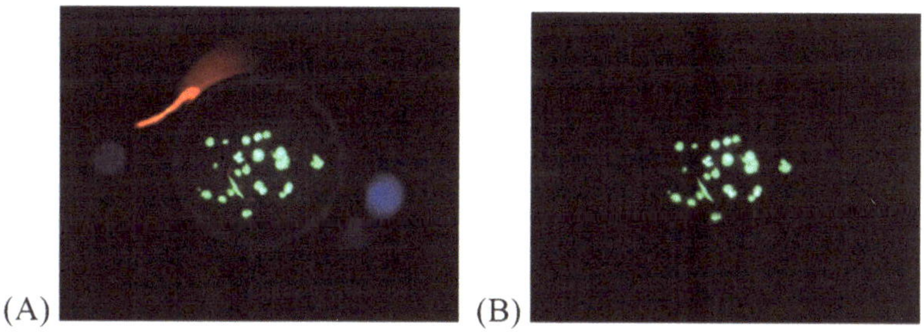

Fig. 7.5. The performance of hue filtering: (**a**) Original image with cells (in *green*), dust (in *red*) and artifacts (in *blue*) and (**b**) Hue filtered image with cells alone. *Green dots* represent live cells on the micropillar chip

7.3.2.2 Background Subtraction, Brightness Filter, and Region of Interest (ROI)

When the hue of the stained cells is similar to background fluorescence, dust, and artifacts, we can use several methods to eliminate unwanted noise fluorescence. Brightness filter (Fig. 7.3) simply reduces pixel brightness values below a certain threshold to zero (Fig. 7.6). This method is particularly useful for images with significant base fluorescence. Background subtraction (Figs. 7.3 and 7.7) can be used as an alternative of, or in conjunction with, the brightness filter in reducing background fluorescence via the rolling ball algorithm [5]. Often times, the undesirable fluorescence signals can be

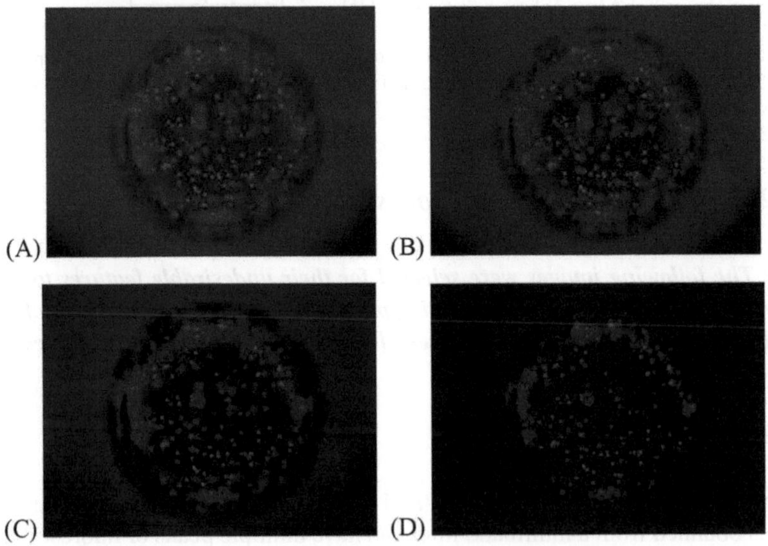

Fig. 7.6. The effect of brightness filter: (**a**) Original image with no background filter applied, (**b**) Brightness filter 10 applied, (**c**) Brightness filter 30 applied, and (**d**) Brightness filter 50 applied. *Blue dots* represent nucleus stained with Hoechst 33342

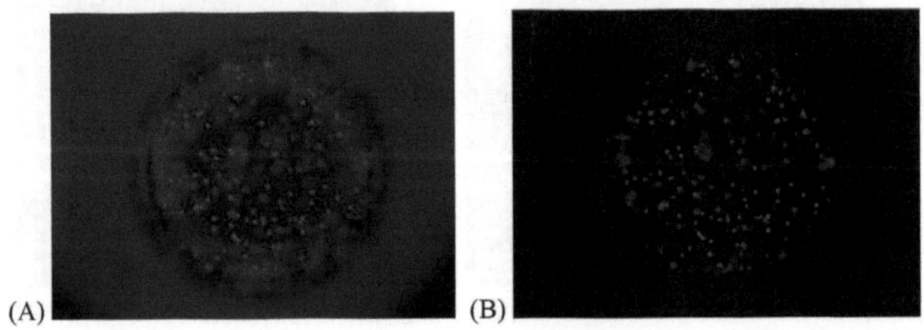

Fig. 7.7. The effect of background subtraction: (**a**) Original image with no rolling background subtraction applied and (**b**) Rolling background subtraction applied

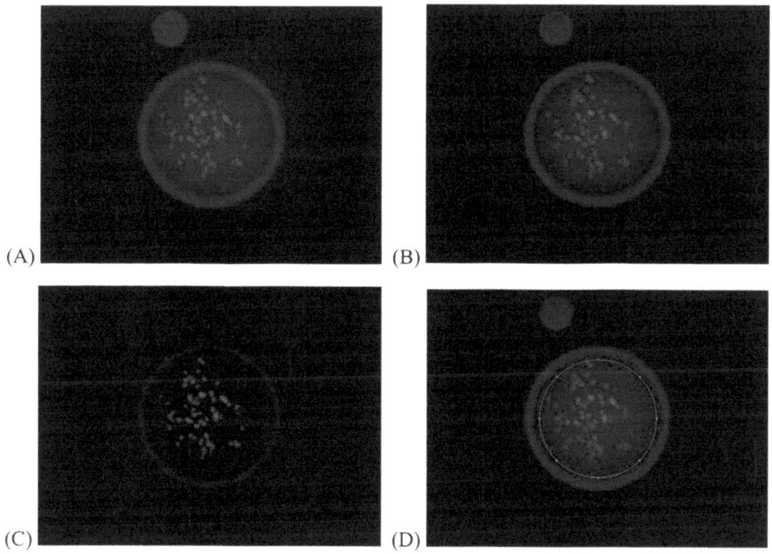

Fig. 7.8. Typical image processing: (**a**) Original image with edge artifacts, (**b**) Brightness filtered image, (**c**) Background subtracted image, and (**d**) Region of interest (ROI) for fluorescence extraction. *Blue dots* represent nucleus stained with Hoechst 33342

found outside the area of the micropillar/microwell where cells are seeded and grown. In that case, we can specify a region of interest (ROI) and have the software only extract fluorescence data from those regions (Fig. 7.8).

7.3.2.3 Outlier Exclusion

When attempting to derive meaningful information from fluorescence data (e.g., dose response curves), a few images can highly skew the results. For example, spot detachment will lead to no fluorescence on certain images whereas smearing or dust can lead to extremely high fluorescence on others. The burden, of course, lies on the scientist to modify or fix the experimental procedures, leading to those problems to ensure the reliability of the fluorescence data. In the meanwhile, there is an option to exclude erroneous fluorescence data beyond two standard deviations of the mean to discount invalid data points (Fig. 7.9). If the variability is expected or part of the experimental design, then this option is not recommended.

7.3.2.4 The Performance of the Plugin

Without any processing, the scanned images often have high background fluorescence and high variability. By applying the various aforementioned features of the plugin, we can significantly reduce the unwanted background fluorescence and

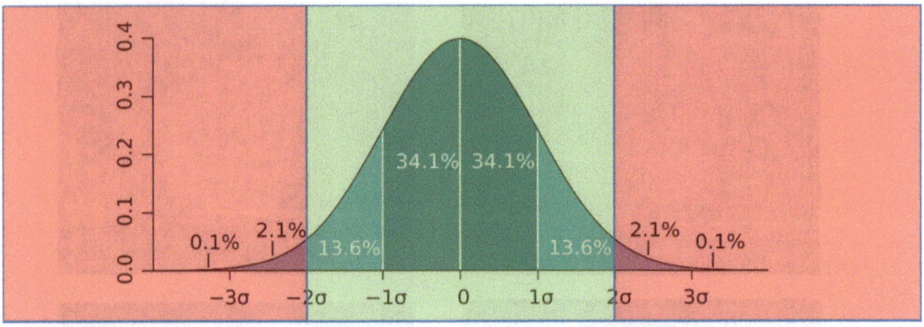

Fig. 7.9. Excluded data points are indicated in *red*

diminish the inflated variability as shown in the following images. Briefly, the micropillar chips were seeded with varying cell concentrations and stained with 2 different dyes—fluo-4 acetoxymethyl ester (AM), and monochlorobimane (mBCl)—to determine the relationship between fluorescence and cell seeding density. Unprocessed raw fluorescence data acquired from the scanned micropillar chips were compared with the processed data (Fig. 7.10). Both Fluo-4 AM and mBCl had high background fluorescence, which was significantly reduced by the plugin. Variability caused by spot detachment, dust, or artifacts was also significantly reduced—for example, see mBCl at 2 million cells/mL.

7.3.3 *Image Deconvolution*

Image processing is a well-developed field, and there are vast and increasing number of solutions that could be borrowed and implemented to enhance signals while eliminating noises. One such method is deconvolution. When cells are encapsulated in a 3D gel, some of them may not be in focus at the same time. As a result, the out-of-focus cell images appear somewhat poor and blurred, which could impact the resulting fluorescence data. Mathematically, blurring can be considered as a convolution between the in-focus (theoretical) image and a blurring factor. The blurring factor could be removed using deconvolution so that an un-blurred image can be remained. Therefore, we are focusing on image deconvolution as a potential solution to the blurring problem. Various image processing toolboxes or suites exist that can perform deconvolution such as the image processing toolbox for MATLAB or the several deconvolution plugins for ImageJ. We decided to use Image Restoration software developed by Advanced Technology Inc. (ATI) for its decoupling and customizability of kernel estimation methods and deconvolution methods. The Kernel estimation method refers to the process of identifying the blurring factor, and the deconvolution algorithm uses the estimated kernel to deconvolve the images. The blind form of Richardson and Lucy (RL) algorithm combined with Inter-Level Intra-Level Deconvolution (ILILD) was determined to be ideal for images of cells

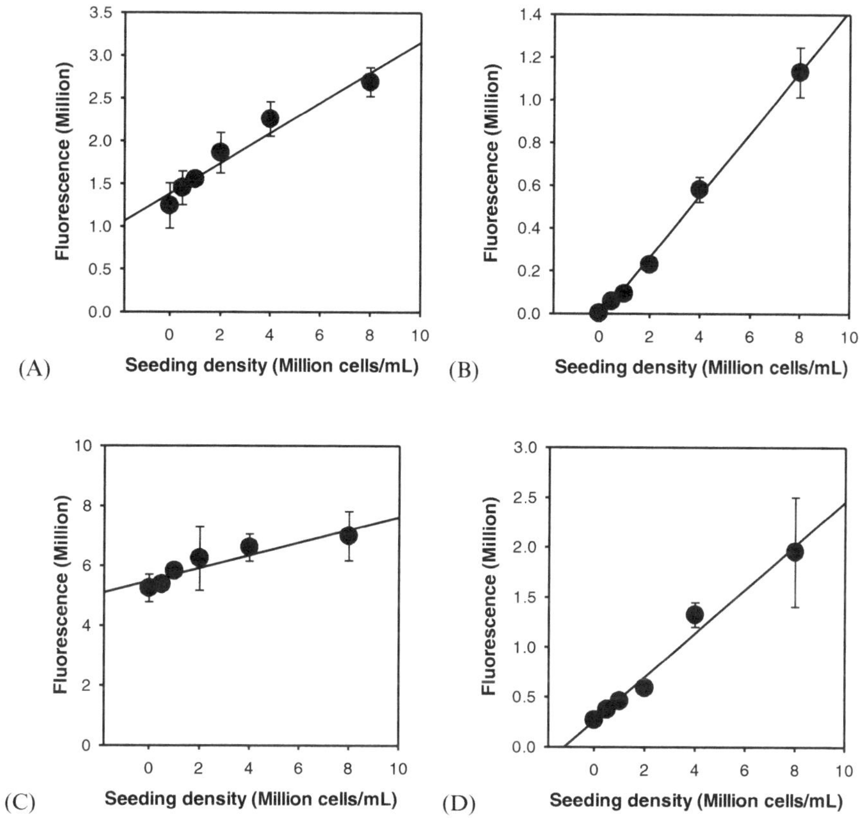

Fig. 7.10. The effect of the BioPrinting plugin (hue filter, background subtraction, brightness filter, ROI, and outlier exclusion) on cell images from the microarray chip: (**a**) Unfiltered data of cells stained with fluo-4 AM, (**b**) Filtered fluo-4 AM data, (**c**) Unfiltered data from cells stained with mBCl, and (**d**) Filtered mBCl data

obtained from the microarray chips. RL is a traditional deconvolution algorithm developed independently by William Richardson and Leon Lucy in 1972 and 1974, respectively. The algorithm was derived from Bayes' theorem, and is an iterative process, the blind form of which alternates between estimating the point spread function (kernel) and the object (deblurred image), and converges on the maximum likelihood solution for the kernel [6]. ILILD is a pyramid structured (multi-layered) algorithm where inter-level deconvolution is performed at multiple levels of resolution, from coarse to fine. In addition, at each level, residual deconvolution (intra-level deconvolution) is performed to recover edge features and details [7].

1. Install and open the Image Restoration software from ATI (Fig. 7.11).
2. Select the 'Kernel estimation' method from the drop-down list. ***Note: RL is recommended.***

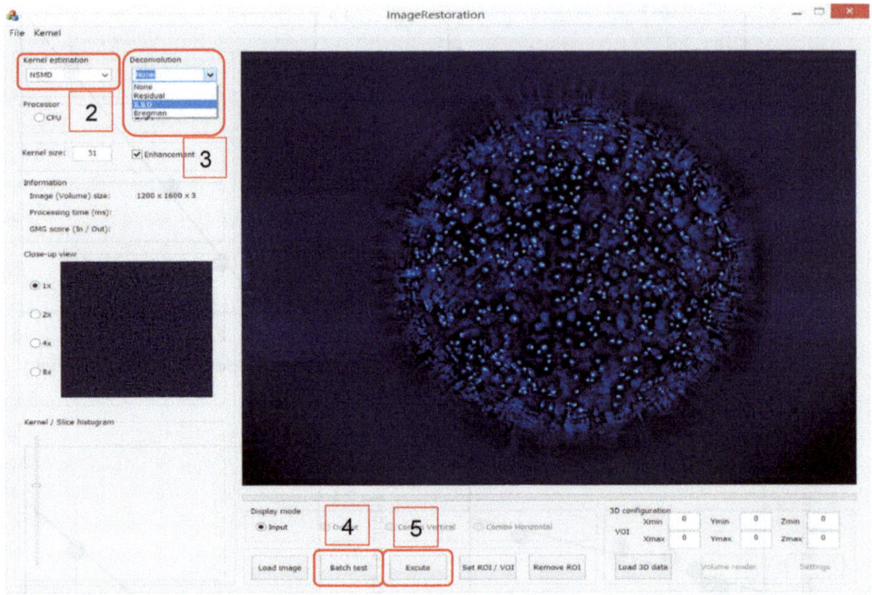

Fig. 7.11. ATI's Image Restoration software for deconvolution. *Blue dots* represent nucleus stained with Hoechst 33342

3. Select the 'Deconvolution' method from the drop-down list. ***Note: ILILD is recommended.***
4. Click on the 'Batch test' button and select the directory where the images are stored.
5. Click 'Execute' to run the software for deconvolution.

7.3.3.1 Performance of Deconvolution

3D cell cultures, especially multi-layered 3D cell cultures in the microwell chip, can be challenging to image due to the cells not being in the same focal plane due to relatively large spotting volume (typically 350–1000 nL). To test the efficacy of deconvolution on multi-layered 3D cell cultures, microwell chips were seeded with a double layer of hoechst 33342-stained cells in varying concentrations, imaged, and processed using the ImageJ Bioprinting plugin as well as the deconvolution software. Deconvolution was found to increase the sensitivity and the slope of the fluorescence calibration as well as the wellness of fit (Fig. 7.12).

7.3.4 Plotting Dose Response Curves with S+ Chip Analysis

1. Open S+ Chip Analysis program by double clicking the S+ Chip Analysis icon on the desktop (Fig. 7.13).
2. Click the 'Experiment' button on the 'Menu' bar to open the 'Experiment condition' window (Fig. 7.14).

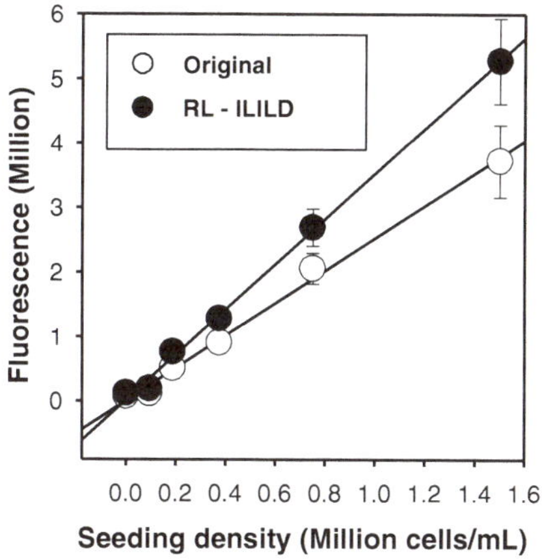

Fig. 7.12. The effect of deconvolution on fluorescence calibration curves. The blind Richardson and Lucy (RL) kernel estimation method and the Inter-Level Intra-Level Deconvolution (ILILD) algorithm were applied to the images of double-layered Hoechst 33342-stained cells

Fig. 7.13. S+ Chip Analysis software icon

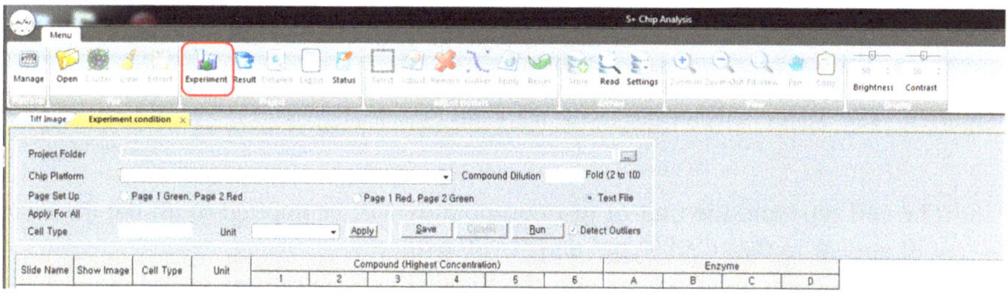

Fig. 7.14. S+ Chip Analysis window with the 'Menu' bar and the 'Experiment condition' window

3. Click the 'Browse' button in the 'Experiment condition' window and select the experiment folder where image data are saved (Fig. 7.15).
4. Click the 'Show' button to see the data. The data will be displayed on a popup window (Fig. 7.15).
5. Select the 'Chip Platform' for the layout of spots (i.e., experimental conditions) on the chip from the drop down list (Fig. 7.16).
6. Enter the dilution factor in the 'Compound Dilution' field (Fig. 7.16).
7. Enter the name of the cells used in the 'Cell Type' field, select the unit of compound measurement (e.g., μM) from drop down list in the 'Unit' field, and then click the 'Apply' button (Fig. 7.16).

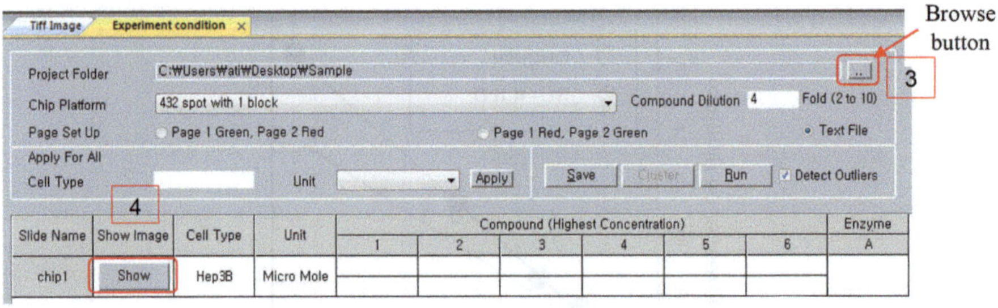

Fig. 7.15. The 'Experiment condition' window with browse button for selecting the desired folder and the 'Show' button for displaying the data, both highlighted in *red*

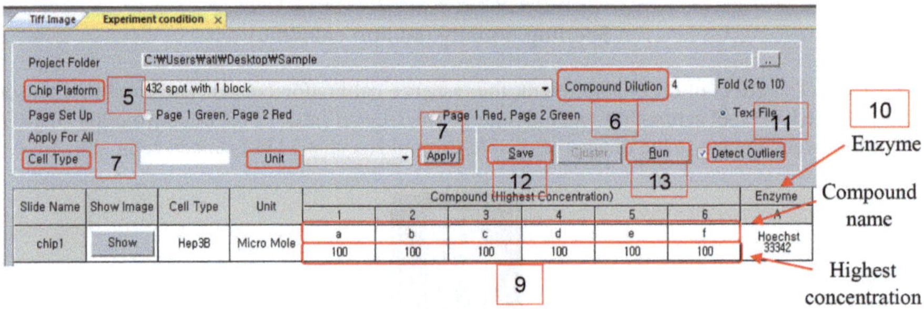

Fig. 7.16. The display of the 'Experiment condition' window

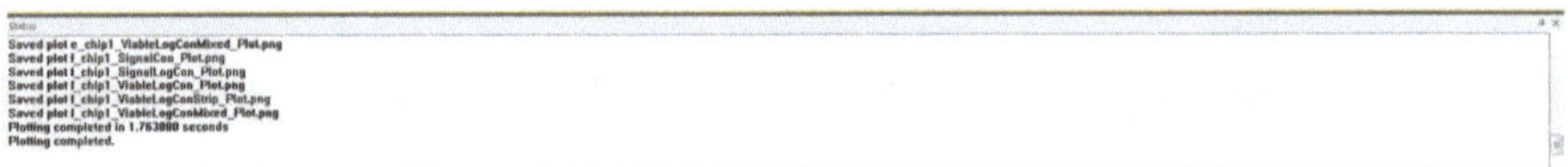

Fig. 7.17. The 'Status' window displaying the status of data analysis

8. The cell type and the unit of test compounds will be applied to all the data and will appear in the 'Cell Type' and 'Unit' grid.
9. Enter the compound names and the highest concentrations tested against the cell type in the first and second row, respectively (Fig. 7.16). ***Note: The name of six compounds and the respective concentrations should be entered individually in column numbered from 1 to 6, respectively.***
10. Enter the name of enzymes or fluorescent dyes or any reagents applied to the chip in the 'Enzyme' grid (Fig. 7.16). ***Note: While analyzing multiple chips, if the compound names, concentration values, and enzymes/fluorescent dye names are identical, the user can copy and paste parameters from one chip to other chips.***
11. If outliers are to be detected and removed automatically, check the 'Detect Outliers' option (Fig. 7.16).
12. After entering the parameters press the 'Save' button to save the parameters (Fig. 7.16).

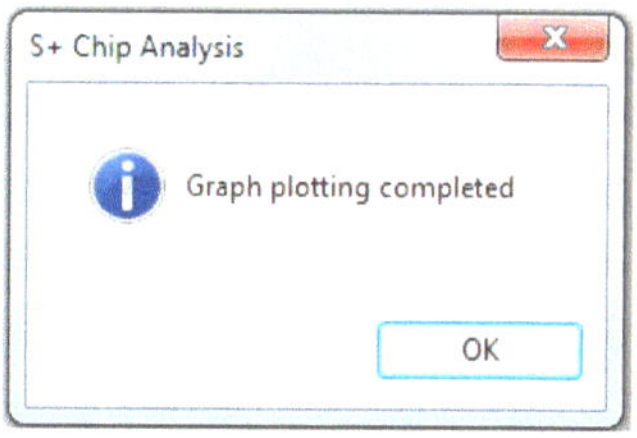

Fig. 7.18. A window displaying 'Graph plotting completed' message

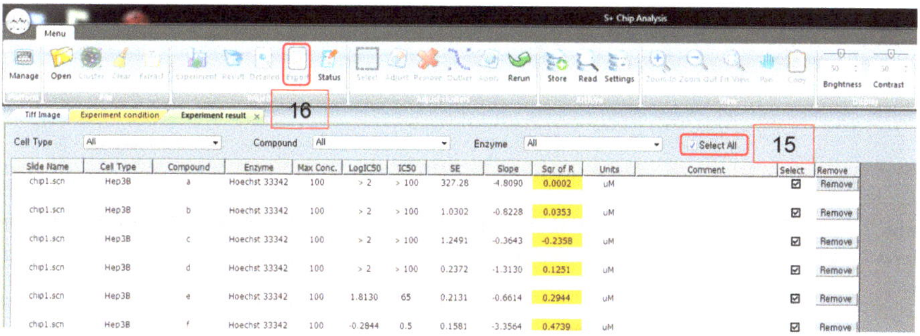

Fig. 7.19. The display of the 'Experiment result' window

13. Press 'Run' button to start analyzing the data (Fig. 7.16). Status of analysis will be displayed in 'Status' window (Fig. 7.17).
14. After the data is analyzed, 'Graph plotting completed' message will appear. Press OK to complete the analysis (Fig. 7.18).
15. The 'Experiment Result' window will show up after the completion of data analysis. Inspect the individual dose response curve and check 'Select All' to select all the graphs obtained or select individual graphs from the chip prior to combining the replicates from different chips (Fig. 7.19).
16. Select 'Export' in the menu bar to export the result/graph into an MS-Word file (Fig. 7.19). ***Note: This process allows one to combine dose response curves from multiple replicate chips and draw one dose response curve.***
17. Once the exporting process is complete, a 'Save As' window will appear with 'Report.doc' as its default file name. Press save if you want to save it with the default file name or else enter a desired file name and press 'Save' button (Fig. 7.20).
18. A 'Select Plot Types' window will show up, check '% Viability vs. Log concentration—summary plot' and click 'Export' button (Fig. 7.21).
19. A pop up window with a message 'Document file created' will appear. Press OK to finish exporting the file (Fig. 7.22).
20. A summary table with test conditions and IC_{50} values as well as corresponding dose response curves will be saved in the same folder containing data file (Fig. 7.23). ***Note: To produce a conventional sigmoidal dose-response curve, the fluorescence intensities of all cell spots are normalized to the fluorescence intensity of 100 % live cell spots (e.g., cell spots contacted with no compound)***

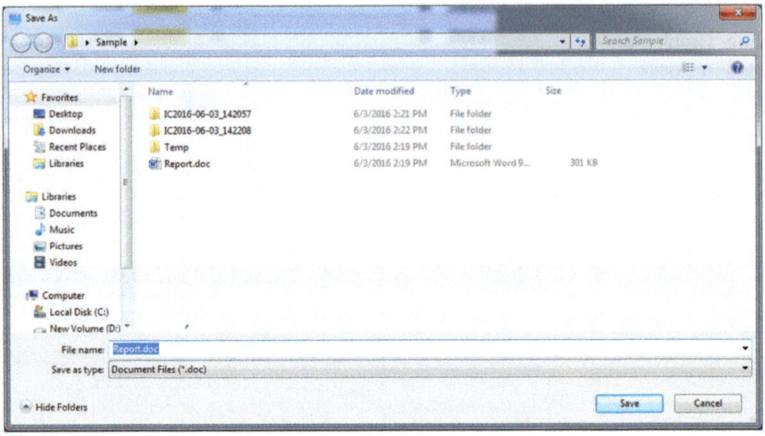

Fig. 7.20. The 'Save As' window appears to save the result in a desired filename inside a desired folder

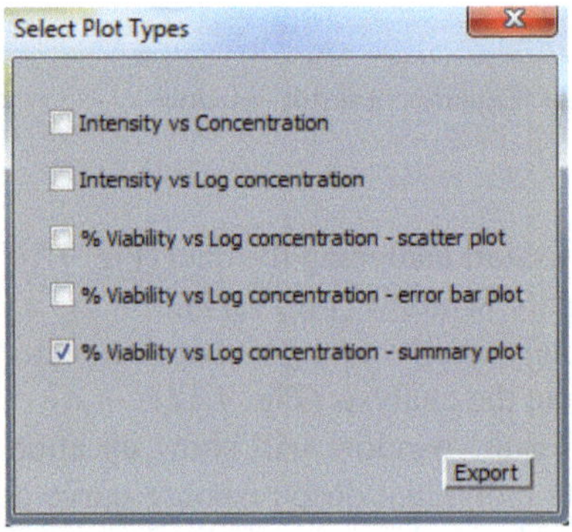

Fig. 7.21. 'Select Plot Types' window

Fig. 7.22. A pop up window displaying 'Document file created' message

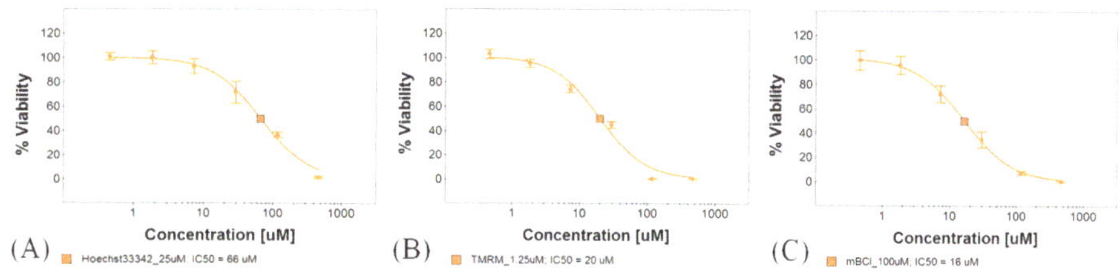

Fig. 7.23. Examples of dose-response curves with IC_{50} values obtained from data analysis in S+ Chip Analysis software. Dose-response curves of lovastatin for (**a**) nucleus morphology, (**b**) mitochondrial membrane potential, and (**c**) glutathione level are presented

and then plotted against the logarithm of test compound concentrations. The sigmoidal dose-response curves (variable slope) and IC_{50} values for each test condition are obtained using the following equation:

$$\mathrm{Y} = \mathrm{Bottom} + \left[\left(Top - \mathrm{Bottom} \right) / \left(1 + 10^{(\mathrm{LogIC_{50}} - \mathrm{X})*\mathrm{H}} \right) \right]$$

where IC_{50} is the midpoint of the curve, H is the hill slope, X is the logarithm of test concentration, and Y is the response (% live cells), starting at Bottom and going to Top with a sigmoid shape.

7.4 Summary

In this chapter, we introduced protocols for processing the images of 3D-cultured cells on the micropillar/microwell chip in a high-throughput manner using ImageJ. We also demonstrated a deconvolution algorithm for deblurring the out of focus images using Image Restoration software developed by Advanced Technology Inc. (ATI), South Korea. Finally, we explained step-by-step protocols for plotting dose-response curves and calculating IC_{50} values using S+ Chip Analysis software, specifically designed for visualizing and analyzing images from microarray chip platforms.

References

1. Lang, P., Yeow, K., Nichols, A., & Scheer, A. (2006). Cellular imaging in drug discovery. *Nature Reviews. Drug Discovery, 5*(4), 343–356. doi:10.1053/j.gastro.2006.06.028.
2. Jahr, W., Schmid, B., Schmied, C., Fahrbach, F. O., & Huisken, J. (2015). Hyperspectral light sheet microscopy. *Nature Communications, 6*, 1–7. doi:10.1038/ncomms8990.
3. Scherf, N., & Huisken, J. (2015). The smart and gentle microscope. *Nature Biotechnology, 33*(8), 815–818. doi:10.1038/nbt.3310.

4. Eliceiri, K. W., Berthold, M. R., Goldberg, I. G., Ibáñez, L., Manjunath, B. S., Martone, M. E., et al. (2012). Biological imaging software tools. *Nature Methods, 9*(7), 697–710. doi:10.1038/nmeth.2084.
5. Sternberg, S. (1983). Biomedical image processing. *IEEE Computer, 16*(1), 22–34.
6. Fish, D. A., Brinicombe, A. M., Pike, E. R., & Walker, J. G. (1995). Blind deconvolution by means of the Richardson–Lucy algorithm. *Journal of the Optical Society of America. A, 12*(1), 58–65.
7. Y. Ding, I. Park, X. Cui, Van Huan. Nguyen, Hakil Kim, Trung Dung Do et al. (2015). Inter-level and intra-level deconvolution based image deblurring algorithm for wide field microscopy. In *2015 8th International Conference on Biomedical Engineering and Informatics (BMEI)*. Shenyang: IEEE (pp. 90–95). doi:10.1109/BMEI.2015.7401479.

Chapter 8
Applications of Microarray Bioprinting

Alexander Roth, Emily Serbinowski, and Moo-Yeal Lee

Contents

8.1 Introduction 161
8.2 Assay Development for Microarray Bioprinting Technologies 162
8.2.1 Hepatotoxicity Assays 162
8.2.1.1 Phase I and Phase II Drug Metabolizing Enzyme Assays 164
8.2.1.2 Drug Transporter Assays 165
8.2.1.3 Oxidative Stress Assays 166
8.2.2 Neurotoxicity Assays 166
8.2.2.1 Oxidative Stress and Related Assays 167
8.2.2.2 Ion Channel Assays 167
8.2.2.3 Drug Metabolism Assays 168
8.3 Simulation of the *In Vivo* Microenvironment: Liver Applications 169
8.4 Other Applications 171
8.5 Summary 171
References 172

8.1 Introduction

Applications for microarray bioprinting include compound toxicity and efficacy assessments, disease diagnostics, and simulation of tissue microenvironments on a small scale. The low volume, high-throughput nature of this system makes it ideal for rapid screening and assay development, particularly in regards towards assays that can be used on various tissue engineering platforms. In this section, we highlight the development of assays for high-throughput screening (HTS) using microarray bioprinting technologies, with a specific focus on toxicity assays that affect the liver and the central nervous system (CNS). Next, we will highlight the various organ systems that can be recapitulated using 3D bioprinting, and the advantages and disadvantages of tissue miniaturization for microarray technologies. Finally, we'll discuss other applications, including tissue regeneration studies, and drug patterning.

M.-Y. Lee, *Microarray Bioprinting Technology*,
DOI 10.1007/978-3-319-46805-1_8

8.2 Assay Development for Microarray Bioprinting Technologies

Adverse drug reactions (ADRs) are among the five leading causes of death in the United States [1]. Among ADRs, hepatotoxicity and cardiotoxicity are the most common events observed. Other organ-specific ADRs, including neurotoxicity and renal toxicity, are observed in the presence of certain drugs. Additionally, the cost for successful drug development can range from $160 million to $2.5 billion, and it takes ten to 15 years from lead compound discovery to clinical evaluation [2]. About 60 % of this cost comes from clinical trials [3]. With roughly a post-marketing failure rate of one drug per year, the human, financial, and capital tolls are further increased.

One of the major issues with drug development is the reliance on animal models to predict ADRs before clinical trials are conducted. Animal models have been shown to be poor predictors for ADRs, as other mammals do not adequately mimic all the human tissues and organ functions [4–7]. Additionally, animal models are generally significantly more expensive than preliminary *in vitro* toxicity studies. Thus, it is a goal of scientists to develop a suite of *in vitro* toxicity assays that can be used to accurately predict human *in vivo* toxicity, and potentially reduce the costs associated with drug development.

In this section, we will focus on assays that can be implemented for *in vitro* toxicity studies on the microarray printing platform. Specific focus will be placed on hepatotoxicity assays, focusing on phase I and phase II drug metabolizing enzymes (DMEs), transporter assays, and oxidative stress detection. Additionally, neurotoxicity will be explored in regards to assays that detect specific mechanisms for neurotoxicity, and how this toxicity is differentiated between adult neuronal cells and neural stem cells (NSCs).

8.2.1 Hepatotoxicity Assays

The liver is arguably the most important organ tied to toxicity assays as the liver is responsible for clearance of toxic substances from the body, and hepatotoxicity is the most common form of ADR that is associated with market withdrawals. Toxicity in the liver can stem from general elevated exposures to certain compounds, known as intrinsic drug-induced liver injury (DILI), or it can stem from individual variations in the tissue microenvironment and the genetic of the individual, known as idiosyncratic drug-induced liver injury (IDILI). Here, we will focus on the assays associated with detection of DILI.

Drug metabolism in the liver is illustrated in Fig. 8.1. An external, apical membrane transporter can allow drugs to enter the cell, though it is also possible for the structure of the molecule to pass through the membrane via diffusion. These compounds are then covalently modified in 2–3 steps to be transported into the bile via

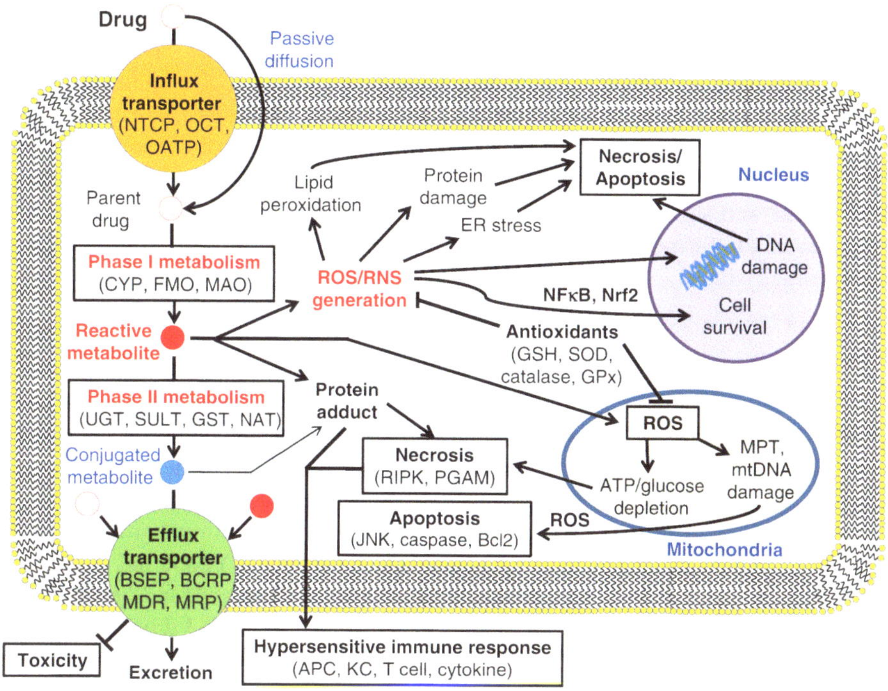

Fig. 8.1. Simplified mechanisms of metabolism-mediated drug toxicity in liver cells. Abbreviations are used as follows: *NTCP* Na^+-taurocholate cotransporting polypeptid, *OCT* organic cation transporter, *OATP* organic anion transporting polypeptide, *BSEP* bile salt export pump, *BCRP* breast cancer resistance protein, *MDR* multidrug resistance protein, *MRP* multidrug resistance-associated protein, *CYP* cytochrome P450, *FMO* flavin-containing monooxygenase, *MAO* monoamine oxidase, *UGT* UDP-glucuronosyltransferases, *SULT* sulfotransferase, *GST* glutathione S-transferase, *NAT* N-acetyl transferase, *ROS* reactive oxygen species, *RNS* reactive nitrogen species, *NFκB* nuclear factor kappa-light-chain-enhancer of activated B cells, *Nrf2* nuclear factor erythroid 2-related factor 2, *GSH* glutathione, *SOD* superoxide dismutase, *GPx* glutathione peroxidase, *MPT* mitochondrial pore transition, *mtDNA* mitochondrial DNA, *RIPK* receptor-interacting serine/threonine protein kinase, *PGAM* phosphoglycerate mutase, *JNK* c-Jun N-terminal kinase, *Bcl2* B-cell lymphoma 2, *APC* antigen-presenting cell, *KC* Kupffer cell

basolateral membrane proteins for eventual elimination. The modification of the structure of the parent drug involves the formation of reactive functional groups such as alcohols and aldehydes (phase I drug metabolism), followed by subsequent reaction with bulkier, less reactive functional groups, such as acetyl, sulfate, and glucuronyl functional groups (phase II drug metabolism). The parent drug, reactive intermediates, and final products can all be transported out of the cell. ADRs in the liver can result from the dysfunction of any DMEs or specific drug transporters, which can have effects ranging from localized inflammation to complete liver failure.

In addition, all forms of the drug may produce reactive byproducts in the form of reactive oxygen species (ROS) and reactive nitrogen species (RNS), which can have the effects of damaging DNA, proteins, and lipids. Mechanisms that clear ROS and RNS from the body can contribute to limiting the effects of ADRs, but if the activities of the enzymes that combat oxidative stress are compromised, this can also have a deleterious effect on hepatocytes. Finally, molecules damaged by any forms of the drug or ROS/RNS can be presented on the surface of the hepatocyte, potentially triggering an acute inflammatory response. While most responses are immediate and short term, sometimes the inflammatory response is sustained and this can lead to further damage, or potentially autoimmunity against the liver, which is potentially fatal.

8.2.1.1 Phase I and Phase II Drug Metabolizing Enzyme Assays

Assays that focus on detection of hepatotoxicity rely on metabolic profiling associated with various DMEs and potential interactions with drugs. The enzyme that is responsible for the majority of the phase I drug metabolism in the liver is cytochrome P450 oxidase (CYP450). CYP450 reacts O_2 with functional groups on parent drugs to produce alcohols, using NADPH as a cofactor. CYP450 has many well studied isoforms that are associated with the metabolism of compounds with significantly varied structures. While other enzymes involved in phase I drug metabolism exist within hepatocytes, polymorphisms in various CYP450s have resulted in the ADRs being observed in the human liver.

CYP450 activity for most clinically relevant isoforms has been determined for both *ex vivo* and *in vitro* studies using coumarin-based metabolism. Each of the CYP450 isoforms can metabolize variants of the fluorogenic substrate of coumarin. For the CYP2B6, CYP2E1, and CYP3A4 isoforms, this results in the formation of 7-hydroxy-4-trifluoromethyl coumarin (HFC) from different substrates [8, 9]. Previous results have shown that HFC production from these substrates is linear with time, meaning there is good correlation with the enzymatic activity. All of the reactions utilize hydroxylation of the ether oxygen that is conjugated to the functional group. Zarowna-Dabrowska *et al.* have alternatively exploited the fluorescent properties of resorufin to detect CYP1A1/2 activity via a 7-ethoresorufin-*O*-deethylase (EROD) assay under confocal microscopy [10]. CYP450 activity detection isn't strictly limited to using direct fluorescent and fluorogenic substrates. Meli *et al.* have demonstrated the use of immunofluorescent microarrays for the detection of CYP2C9 and CYP3A4 activity [11].

The advantage of fluorogenic substrates is that they can be both printed in microarray chip platforms, and detected under fluorescence microscopy. Generally, fluorogenic substrates are also more sensitive over a broader concentration range, and can hence detect both low and high activities of CYP450s quite easily. Promega provides several commercially available assays that can be utilized within the microarray chip platforms for detection of phase I metabolic activity, in fluorescent, colorimetric, and luminescent forms, through fluorescence is most compatible with cellular microarrays. The drawback to these substrates is that if the cell line is capable of significant phase II metabolism, inhibitors need to be added to the solution to properly assess CYP450 activity. Additionally, there are substrates that exist which can be metabolized by isolated CYP450s, but may not be adequate to permeate the cell membrane. Human liver microsomes (HLMs) may be used to convey or alter CYP450 activity to cells lacking CYP450 activity, or can be used as a suitable control for assaying activity of certain compounds with various CYP450s [9]. HLM activity can be assayed using microarray bioprinting technology, which offers the alternative of conventional 96-well plate assays at a lower cost.

Phase II drug metabolism is the replacement of reactive functional groups added by phase I drug metabolism with bulkier, easily cleared functional groups. Phase II enzymes include uridine 5′-diphosphoglucuronosyltransferases (UGTs), sulfotransferases (SULTs), glutathione S-transferases (GSTs), and N-acetyltransferases (NATs). For all of these enzymes, there exist clinically relevant polymorphisms that make certain individuals more susceptible to ADRs. Specific protein assays that are relevant to ADRs have been developed for measuring UGT and GST activity. UGT activity can be measured by using 4-methylumbelliferone (4-MU), a fluorescent substrate of the coumarin family that undergoes glucuronidation in the presence of UGT [12, 13]. Additionally, owing to the role of UGT in clearance of steroids, β-estradiol can be used to assay UGT1A1 activity via detection of estradiol 3-glucuronide [14]. GST can be measured and distinguished using 1-chloro-2,4-dinitrobenzene (CDNB) as a substrate [15]. This reaction involves the removal of the chlorine via reaction with the sulfur on glutathione (GSH). While CDNB can be used as a substrate for total GST activity, it is measured via absorbance, and less ideal for microarray bioprinting applications. Monochlorobimane (MCB) is recommended as the reaction with GSH produces a fluorescent compound (excitation 380 nm/emission 461 nm), which has been used previously to assess GSH levels in cells.

8.2.1.2 Drug Transporter Assays

Transporters represent a very diverse class of enzymes. Influx transporters and efflux transporters are associated with the transport of various compounds from the lumen (influx) to the bile (efflux). Generally, toxicity can result from dysfunction in efflux transporter, as reactive metabolites can accumulate in the cytoplasm if they cannot escape the cell. Included in the list of efflux transporters are breast cancer resistance protein (BCRP), bile salt export pump (BSEP), multidrug resistance protein (MDR), and multidrug resistance-associated protein (MRP). BCRP, BSEP, and certain MDRs and MRPs have all been linked with particular ADRs if their function is compromised.

BSEP is the most common target for transporter activity measurement owing to its altered function during cholestasis. Assays used for measuring BSEP activity include the vesicular transport assay (VTA) and the ATPase assay [16]. Both assays utilize ATP as a co-transporter molecular necessary to drive metabolites out of the cell against concentration gradients. While this is specialized for the purpose of measuring BSEP activity, this method can be used for measuring ABC transporter activity. In addition to these assays, drug uptake assays are available for various metabolic compounds [17]. MDR1 has a significant amount of substrates available that have similar substrate specificity with CYP3A4, owing to the co-regulation of both proteins by the pregnane X receptor (PXR) transcription factor [18–20]. Induced MDR1 activity can be measured by rifampicin, while inhibition can be measured verapamil [20]. Additionally, MRP2 can be inhibited by 5(6)-carboxy-2′,7′-dichlorofluorescein (CDCF), which is a commercially available fluorogenic substrate [14, 21, 22]. While transporter detection is possible with fluorescent and fluorogenic substrates in the standard well-plate format, only genetic expression of transporters has been achieved in cellular microarrays.

8.2.1.3 Oxidative Stress Assays

While oxidative stress can affect numerous organelles, antioxidant enzymes localized to the mitochondria are the antioxidants most associated with individualized ADRs. This makes the mitochondrion the primary indicator for liver functionality tests related to oxidative stress. For general mitochondrial toxicity assays, tetramethyl rhodamine (TMRM) can be used to stain for measuring mitochondrial membrane potential, as depolarization of the membrane is a symptom of apoptosis [23]. Cytochrome c release and oxygen consumption rate (OCR) may also be used to determine effects on energetic metabolism [24]. OCR determination utilizes the metabolic turnover of O_2 in energetic metabolism [25, 26]. OCR can be measured fluorescently using a variety of methods, though fluorescent probes are most common. MitoXpress from Luxcel (ex/em 380/650) can be used for direct fluorescence measurement of OCR by quantifying oxidative phosphorylation, though time-resolved fluorescence (TR-F) is recommended instead of direct fluorescence. Similar assays are available for OCR measurements from other companies.

If antioxidant levels are innately low, GSH may also be used to determine effects of drug on mitochondria. MCB may be used as a fluorogenic substrate to quantify GSH levels [27], but the activity of GSH as a function of antioxidant stress must be distinguished from that of GSH used in phase II drug metabolism. While GSH acts as a general indicator for oxidative stress, this can be sufficient enough to indicate dysfunction with antioxidant enzymes, or even the general presence of an ADR.

8.2.2 Neurotoxicity Assays

Aside from hepatotoxicity, ADRs are also found in neural cell lineages, particularly neural stem cells (NSCs) [28]. NSCs are the cells that eventually differentiate into the various lineages associated with the central nervous system, including neurons, oligodendrocytes, and astrocytes. Much of the toxicity that drives ADRs in NSCs relates to damaging or blocking ion channels on the cell membrane surface, inducing oxidative stress, denaturing of nucleic acids and proteins from drug metabolites, membrane compromise, and both caspase-dependent and -independent apoptosis. This section will deal with assays that detect these various mechanisms, focusing on assays that can be used in microarray bioprinting applications.

8.2.2.1 Oxidative Stress and Related Assays

Like in hepatocytes, NSCs may undergo apoptosis in the presence of certain compounds. Apoptosis may be derived from either direct interaction of certain compounds with molecules within the pathways necessary for apoptosis, or it can derive from oxidative stress related to reactive metabolites. In order to detect this behavior, assays have focused on observing the behavior of specific molecules within these pathways. Some of these methods, such as simple live-dead stains discussed in Chap. 5 can elucidate the basic mechanisms of toxicity, pointing towards mitochondrial dysfunction, glutathione depletion, or caspase-mediated apoptosis. In addition, oxidative stress often couples with reduced membrane integrity, and the potential for nucleic acid and peptide conjugation from reactive metabolites. While these cellular symptoms aren't necessarily indicative of individualized ADRs, they are all indicative of generalized ADRs.

Fluorescent methods used to quantify these mechanisms are more advantageous for microarray bioprinting technologies owing to the small volume dispensed in the microarray, and thus the increased sensitivity necessary to detect the potential for ADRs. For protein and nucleic acid modification, benzo(a)pyrene diol epoxide (BPDE) is used in conjunction with a colorimetric enzyme-linked immunosorbent assay (ELISA) to detect such molecules that may have been conjugated due to drug byproducts. The commercially available ELISA quantifies adduction via horseradish peroxidase (HRP) conjugated to secondary antibodies. However, this secondary antibody could potentially be replaced by a fluorescent one to make it suitable for microarray platforms [29, 30]. Abcam provides a kit (ab112158, ex/em: 490/520 nm) that can quantify adduction and GSH levels via measuring thiol groups with their reagent as it conjugates with free thiols on proteins and GSH.

For cell membrane compromise, lactate dehydrogenase (LDH) can be used to quantify the viability of cells in 96-well plates and in microwell chips [31–33]. When cells lyse or the cellular membrane is compromised, LDH may leak into the surrounding environment. CytoTox One has manufactured an assay that utilizes the conversion of lactate to pyruvate via NAD^+ as a cofactor for NADH. It then quantifies the level of NADH via NADH reductase, which converts resazurin to resorufin, a fluorescent compound.

8.2.2.2 Ion Channel Assays

Another way ADRs are triggered in NSCs involve the blocking of ion channels. Ion channels are associated with maintaining plasma membrane potential, but they can also be found on the membranes of various organelles. Within axonal differentiated

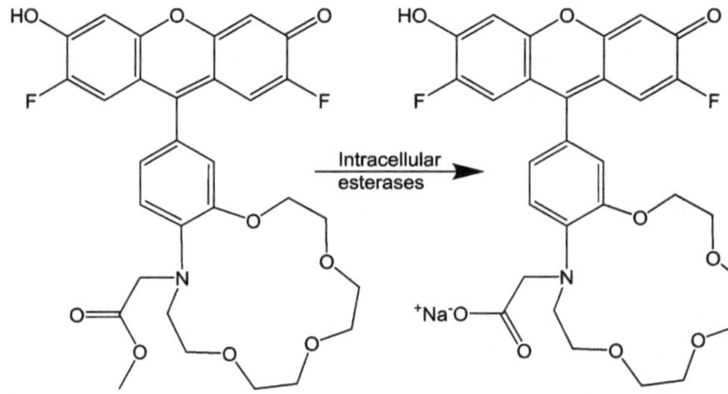

Fig. 8.2. The conversion of CoroNa green via intracellular esterases to a fluorescent indicator for intracellular Na^+ ions

neural cell lineages and NSCs, ion channels also play a role in synaptic transmission and the generation of action potentials, which is a common behavior associated with cells of the CNS and all versions of muscle cells. Disruption in ion channels may lead to disruption in the transmittance of action potentials, and the subsequent loss of signals. K^+, Na^+, Ca^{2+}, and Cl^- may all be blocked within neural cells and NSCs. The disruption in ion flow could result from competitive ionic interactions of ions with similar valence charges (i.e., transition elements Ni^{2+} and Cd^{2+} with Ca^{2+}), covalent modification of proteins, or competitive binding with necessary ligands that may allow channels to open and close. Common compounds used as ion channel blocker controls include 4-aminopyridine, disopyramide, amlodipidine besylate, and 4,4′-diisothiocyanostilbene-2,2′-disulfonic acid (DIDS).

There are commercially available assays used to detect ion channel activity and ion channel blocking. FluxOR potassium ion channel assay from Life Technologies uses thallium ions to trigger the opening of K^+ channels, activating the fluorescent dye inside the cell, while also using probenecid to prevent the exit of the fluorescent dye. The CoroNa green sodium indicator from Life Technologies uses a fluorescein based molecule with a binding pocket for Na^+, which is activated inside cells despite Na^+ ion concentrations generally being greater extracellularly (Fig. 8.2). Fluo-4AM reagent from Life Technologies is believed to detect Ca^{2+} molecules intracellularly via cleavage of the acetoxymethyl ester groups. This may be coupled with probenecid to maintain the fluorescent moieties inside the cell. Premo halide sensor from Life Technologies uses I^- to mimic the influx of Cl^- ions. I^- will quench a yellow fluorescent agent inside the cell. It is known so far that Fluo-4AM is compatible with a high-throughput bioprinting system, and that the rest of these reagents are cell compatible.

8.2.2.3 Drug Metabolism Assays

Another such mechanism for detecting NSC toxicity is through drug metabolism itself. Like hepatocytes, adult neural cells and NSCs metabolize drugs and can be adversely affected by mutations in DMEs [34]. Drug metabolism in NSCs utilizes a similar process of phase I and phase II metabolism, and there are separate classes of transporters for compound influx and efflux. CYP450s are the DMEs primarily responsible for phase I drug metabolism in NSCs, and there is some overlap with the individual CYP450 isoforms that are responsible for drug metabolism. Additionally, phase II DME of the GST family have been tied to individual ADRs, and dysfunction in transporter MDR1 may also increase susceptibility to ADRs.

NSC assays which deal with drug metabolism focus on similar assays that are involved in assessing DME-related toxicity in hepatocytes. While the protocols are identical for detection of activity with these enzymes, many of the CYP450 isoforms unique to NSCs are involved in the metabolism of fatty acids and steroids [35, 36]. Thus, substrates for metabolism and assaying activity of neuronal CYP450 enzymes can be such biological compounds, or their analogues. Likewise, many of the GST and MDR1 assays used on hepatocytes are also compatible with NSCs, including CDNB and MCB for GST activity, and verapamil for MDR1. These assays are compatible with cells and are fluorescent, making them suitable assays to be used for determining toxicity on a microarray chip platform.

8.3 Simulation of the *In Vivo* Microenvironment: Liver Applications

Automatic liquid dispensation for microarray technologies has until recently seen limited use outside the organism. While research has focused on development of whole 3D tissue constructs, technology has not reached the point where whole organs can be printed for transplantation purposes [37]. The creation of physiological networks is quite complex, and maintaining and mimicking all of the necessary functions of a 3D-bioprinted tissue is difficult. However, microarray 3D bioprinting can be used to mimic small scale *in vivo* behaviors. While this does not make it suitable for large scale cultures, miniaturization of the bioprinting process can still give insight into certain *in vivo* behaviors in a high-throughput manner. In this section, we will focus on the use of bioprinting 3D cell cultures for high-throughput applications, with a focus on co-culture systems related to mimicking the liver (Fig. 8.3).

In order to understand the nature of co-culture systems, we must also understand the cells and vasculature associated with the tissue that is to be mimicked. In the case of livers, there are four types of cells that make up over 95 % of the cell volume and count: hepatocytes, Kupffer cells (KCs), sinusoidal endothelial cells, and hepatic stellate cells [38, 39]. Hepatocytes make up the majority of the cell volume and count (roughly 60 % of the cell count and 80 % of the cell mass) [38]. These cells serve a myriad of functions, including drug metabolism, bile flow, cholesterol synthesis, glycolysis, gluconeogenesis, and amino acid catabolism [39]. The roles the hepatocytes take on relate closely towards their proximity towards the portal vein, the main supplier of oxygen to the liver [39]. The other cell types are localized to other parts of the liver. The KCs are the resident macrophages of the liver that initiate inflammation in response to toxic injury [38, 40, 41]. Sinusoidal endothelial

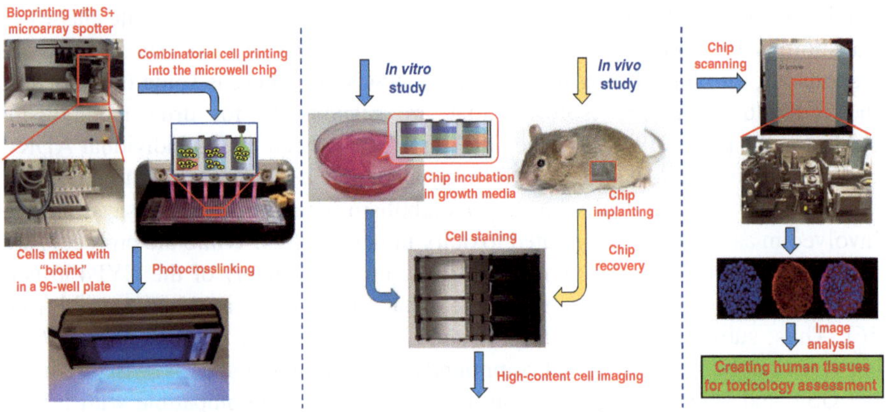

Fig. 8.3. Microarray three-dimensional (3D) bioprinting technology for creating human tissues by dispensing multiple human cells in biomimetic hydrogels ('bioinks') layer-by-layer with precision printing robots, thereby potentially revolutionizing tissue engineering and disease modeling for screening therapeutic drugs and studying toxicology. Miniaturized liver tissue blocks (as small as 1 mm^3) are generated by printing several layers of human hepatic cells in photocrosslinkable hydrogels with extracellular matrices (ECMs) and growth factors (GFs) into a microwell chip using S+ MicroArrayer. After gelation, the microwell chip containing hundreds of biomimetic conditions is incubated in a petri dish with growth media or implanted in immunocompromised animals for rapidly testing optimum microenvironments to create biomimetic human tissues. Created human liver tissue constructs are stained and scanned with S+ Scanner for high-content imagining (HCI) of hepatic functions as well as predictive assessment of drug toxicity

cells border the blood vessels, and are important as they have interactions with any blood cells (including KCs) and can present antigens to CD4+ (mature) helper T cells [42, 43]. They are also important on limiting the effects of shear on hepatocytes [38, 39, 44]. Finally, hepatic stellate cells exist in the space of disse, a region between the endothelial cells, and the hepatocytes. The stellate cells store fatty acids and some vitamins, but they may behave like fibroblasts and synthesize collagen if liver restructuring is necessary [38, 39, 45].

To mimic the behavior of the liver, focus has shifted towards 3D co-culture systems, utilizing both hepatocytes and non-parenchymal cells. While KCs are most frequently incorporated into these systems owing to their role in response towards ADRs, the other cell types have been used too [46, 47]. Model systems that have been used to create this environment include large scale hepatic bioreactor [48, 49], microfluidic technologies [50–52], layered-sheets of fibroblasts to initiate growth conditions [53], and 3D hydrogel technologies for microarrays [54]. For the drug discovery process, including testing for ADRs, 3D hydrogel microarrays are ideal because of their high throughput nature. Hepatic bioreactors cannot be scaled down at low cost, microfluidics can provide some replicates, but not with the same efficiency as a microarray can, and fibroblast systems could potentially be more complicated than the hydrogels, and many really do not adequately mimic the 3D liver microenvironment. Therefore, it is more desirable to use a microarray system for drug development testing.

Currently, several systems are in place to print co-cultures of 3D systems, recapitulate the *in vivo*-like behavior of various liver cell lines on both 2D and 3D platforms, or at the very least co-culture various cell types in a hydrogel platform. Kwon *et al.* have demonstrated the feasibility of using a micropillar chip as a platform for culture of printed cells, and have modified the expression of various DMEs via adenoviral transduction systems to examine individual toxicities [30]. Kang *et al.* established the viability of primary rat hepatocytes (PRHs) co-cultured with sinusoidal endothelial cells in a transwell membrane plate and established the maintenance in the level of expression of CYP450 activity. Bhise *et al.* printed co-cultures of HepG2 and C3A into spheroids for the sake of bioreactor functions on a liver-on-a-chip platform, and retained transporter function. All of these results show promise for the ability of microarray bioprinting towards the recapitulation of the 3D *in vivo*-like behavior.

8.4 Other Applications

The major goal of most moderate to large scale 3D bioprinting is to recapitulate *in vivo* physiology in such a way that the system can be used as a model for various tissues, or even be used to replace tissue. While the microarray system is not suitable for direct implantation, small scale mimics of various tissues can be used to model regeneration and cell growth. Though the early part of this chapter has detailed the printing of cells and compounds for applications in hepatic tissues and neural tissues, microarray bioprinting is not limited to just these organs. Ma *et al.* have printed cell-laden gelatin methacrylate (GelMA) hydrogels to test for regeneration of periodontal tissue in periodontitis [55]. The platform for printing was not a micropillar/microwell chip as discussed in much of the book; rather, it was a flat surface pre-treated with other chemicals to improve gel attachment [55]. Other tissues have also been printed at microarray levels. However, the printing technique can best be described as micropatterning for the development of larger 3D scaffolds. In addition to cell and tissue engineering, scientists have printed hydrogels containing test compounds on micropatterned surfaces to test drug toxicity [56]. While compound-cell interactions have been discussed earlier in this chapter, this feature can be used to examine hydrogels as drug delivery systems on a microarray platform, rather than just for cell and tissue engineering constructs.

8.5 Summary

Microarray bioprinting is a useful tool for miniaturized assessments of *in vivo*-like behavior. It offers a small-scale, high-throughput approach in tissue engineering, disease modeling, and drug development research. Assays for detecting cellular functions and toxicity can be performed to minimize the use of otherwise expensive reagents. While macroscale 3D bioprinting has yet to effectively recapitulate whole organ systems, the 3D printing of hydrogels at the microarray level still gives insight into particular details of organ functions, and cells may still cluster and reorganize themselves in ways which are comparable to *in vivo* structures. It is this miniaturization that allows a high-throughput testing of conditions to optimize treatments for disease targets, determine effective nontoxic drugs, and hopefully lead towards greater discoveries in tissue engineering and tissue regeneration research.

References

1. Reuben, A., Koch, D. G., & Lee, W. M. (2010). Drug-induced acute liver failure: Results of a U.S. multicenter, prospective study. *Hepatology, 52*(6), 2065–2076. doi:10.1002/hep.23937.
2. Morgan, S., Grootendorst, P., Lexchin, J., Cunningham, C., & Greyson, D. (2011). The cost of drug development: A systematic review. *Health Policy (Amsterdam, Netherlands), 100*(1), 4–17. doi:10.1016/j.healthpol.2010.12.002.
3. *Preventable adverse drug reactions: A focus on drug interactions*. 2016. Retrieved from http://www.fda.gov/Drugs/DevelopmentApprovalProces.
4. Elliott, N. T., & Yuan, F. A. N. (2011). A review of three-dimensional in vitro tissue models for drug discovery and transport Studies. *Journal of Pharmaceutical Sciences, 100*(1), 59–74. doi:10.1002/jps.22257.
5. Godoy, P., Hewitt, N. J., Albrecht, U., Andersen, M. E., Ansari, N., Bhattacharya, S., et al. (2013). Recent advances in 2D and 3D in vitro systems using primary hepatocytes, alternative hepatocyte sources and non-parenchymal liver cells and their use in investigating mechanisms of hepatotoxicity, cell signaling and ADME. *Archives of Toxicology, 87*(8), 1315–1530. doi:10.1007/s00204-013-1078-5.
6. Kimlin, L. C., Casagrande, G., & Virador, V. M. (2013). In vitro three-dimensional (3D) models in cancer research: An update. *Molecular Carcinogenesis, 52*(3), 167–182. doi:10.1002/mc.21844.
7. Keogh, J. P. (2012). Membrane transporters in drug development. *Advances in Pharmacology, 63*, 1–42. doi:10.1016/B978-0-12-398339-8.00001-X.
8. Lin, J., Schyschka, L., Mühl-Benninghaus, R., Neumann, J., Hao, L., Nussler, N., et al. (2012). Comparative analysis of phase I and II enzyme activities in 5 hepatic cell lines identifies Huh-7 and HCC-T cells with the highest potential to study drug metabolism. *Archives of Toxicology, 86*(1), 87–95. doi:10.1007/s00204-011-0733-y.
9. Walsky, R. L., & Obach, R. S. (2004). Validated assays for human cytochrome P450 activities. *Drug Metabolism and Disposition, 32*(6), 647–660. doi:10.1124/dmd.32.6.647.
10. Zarowna-Dabrowska, A., McKenna, E. O., Schutte, M. E., Glidle, A., Chen, L., Cuestas-Ayllon, C., et al. (2012). Generation of primary hepatocyte microarrays by piezoelectric printing. *Colloids and Surfaces B: Biointerfaces, 89*(1), 126–132. doi:10.1016/j.colsurfb.2011.09.016.
11. Meli, L., Jordan, E. T., Clark, D. S., Linhardt, R. J., & Dordick, J. S. (2012). Influence of a three-dimensional, microarray environment on human cell culture in drug screening systems. *Biomaterials, 33*(35), 9087–9096. doi:10.1016/j.biomaterials.2012.08.065.
12. Leite, S. B., Iwona, W.-Z., Zaldivar, J. M., Airola, E., Reis-Fernandes, M. A., Mennecozzi, M., et al. (2012). 3D HepaRG model as an attractive tool for toxicity testing. *Toxicological Sciences, 130*(1), 106–116.
13. Leite, S. B., Teixeira, A. P., Miranda, J. P., Tostões, R. M., Clemente, J. J., Sousa, M. F., et al. (2011). Merging bioreactor technology with 3D hepatocyte-fibroblast culturing approaches: Improved in vitro models for toxicological applications. *Toxicology in Vitro, 25*(4), 825–832. doi:10.1016/j.tiv.2011.02.002.

14. Chang, J. H., Plise, E., Cheong, J., Ho, Q., & Lin, M. (2013). Evaluating the in vitro inhibition of UGT1A1, OATP1B1, OATP1B3, MRP2, and BSEP in predicting drug-induced hyperbilirubinemia. *Molecular Pharmaceutics, 10*(8), 3067–3075.
15. Liang, Q., Sheng, Y., Jiang, P., Ji, L., Xia, Y., Min, Y., et al. (2011). The gender-dependent difference of liver GSH antioxidant system in mice and its influence on isoline-induced liver injury. *Toxicology, 280*(1–2), 61–69. doi:10.1016/j.tox.2010.11.010.
16. Kis, E., Ioja, E., Rajnai, Z., Jani, M., Méhn, D., Herédi-Szabó, K., et al. (2012). BSEP inhibition: In vitro screens to assess cholestatic potential of drugs. *Toxicology In Vitro, 26*(8), 1294–1299. doi:10.1016/j.tiv.2011.11.002.
17. Lee, J. K., Marion, T. L., Abe, K., Lim, C., Pollock, G. M., & Brouwer, K. L. R. (2010). Hepatobiliary disposition of troglitazone and metabolites in rat and human sandwich-cultured hepatocytes: Use of Monte Carlo simulations to assess the impact of changes in biliary excretion on troglitazone sulfate accumulation. *The Journal of Pharmacology and Experimental Therapeutics, 332*(1), 26–34. doi:10.1124/jpet.109.156653.tion.
18. Cascorbi, I., & Haenisch, S. (2010). Pharmacogenetics of ATP-binding cassette transporters and clinical implications. *Methods in Molecular Biology, 596*, 95–121. doi:10.1007/978-1-60761-416-6.
19. Fromm, M. F. (2004). Importance of P-glycoprotein at blood-tissue barriers. *Trends in Pharmacological Sciences, 25*(8), 423–429. doi:10.1016/j.tips.2004.06.002.
20. Stieger, B. (2011). The role of the Sodium-Taurocholate Cotransporting Polypeptide (NTCP) and of the Bile Salt Export Pump (BSEP) in physiology and pathophysiology of bile formation. In M. F. Fromm & R. B. Kim (Eds.), *Drug transporters*. Berlin: Springer. doi:10.1007/978-3-642-14541-4.
21. Goral, V. N., Hsieh, Y.-C., Petzold, O. N., Clark, J. S., Yuen, P. K., & Faris, R. A. (2010). Perfusion-based microfluidic device for three-dimensional dynamic primary human hepatocyte cell culture in the absence of biological or synthetic matrices or coagulants. *Lab on a Chip, 10*(24), 3380–3386. doi:10.1039/c0lc00135j.
22. Thompson, R. A., Isin, E. M., Y, L., Weidolf, L., Page, K., Wilson, I., et al. (2012). In vitro approach to assess the potential for risk of idiosyncratic adverse reactions caused by candidate drugs. *Chemical Research in Toxicology, 25*(8), 1616–1632. doi:10.1021/tx300091x.
23. Xu, J. J., Diaz, D., & O'Brien, P. J. (2004). Applications of cytotoxicity assays and pre-lethal mechanistic assays for assessment of human hepatotoxicity potential. *Chemico-Biological Interactions, 150*(1), 115–128. doi:10.1016/j.cbi.2004.09.011.
24. Porceddu, M., Buron, N., Roussel, C., Labbe, G., Fromenty, B., & Borgne-Sanchez, A. (2012). Prediction of liver injury induced by chemicals in human with a multiparametric assay on isolated mouse liver mitochondria. *Toxicologic Pathology, 129*(2), 332–345.
25. Hynes, J., Swiss, R. L., & Will, Y. (2012). High-throughput analysis of mitochondrial oxygen consumption. In C. M. Palmeira & A. J. Moreno (Eds.), *Mitochondrial bioenergetics: Methods and protocols* (Vol. 810, pp. 103–117). New York: Humana Press. doi:10.1007/978-1-61779-382-0.
26. Aleo, M. D., Luo, Y., Swiss, R., Bonin, P. D., Potter, D. M., & Will, Y. (2014). Human drug-induced liver injury severity is highly associated with dual inhibition of liver mitochondrial function and bile salt export pump. *Hepatology, 60*(3), 1015–1022. doi:10.1002/hep.27206.
27. Ong, M. M. K., Latchoumycandane, C., & Boelsterli, U. A. (2007). Troglitazone-induced hepatic necrosis in an animal model of silent genetic mitochondrial abnormalities. *Toxicological Sciences, 97*(1), 205–213. doi:10.1093/toxsci/kfl180.
28. Fernandes, T. G., Kwon, S. J., Bale, S. S., Lee, M. Y., Diogo, M. M., Clark, D. S., et al. (2010). Three-dimensional cell culture microarray for high-throughput studies of stem cell fate. *Biotechnology and Bioengineering, 106*(1), 106–118. doi:10.1002/bit.22661.
29. Fernandes, T. G., Kwon, S. J., Lee, M. Y., Clark, D. S., Cabral, J. M. S., & Dordick, J. S. (2008). On-chip, cell-based microarray immunofluorescence assay for high-throughput analysis of target proteins. *Analytical Chemistry, 80*(17), 6633–6639. doi:10.1021/ac800848j.
30. Kwon, S. J., Lee, D. W., Shah, D. A., Ku, B., Jeon, S. Y., Solanki, K., et al. (2014). High-throughput and combinatorial gene expression on a chip for metabolism-induced toxicology screening. *Nature Communications, 5*, 3739. doi:10.1038/ncomms4739.

31. Ulasov, I., Nandi, S., Dey, M., Sonabend, A. M., & Lesniak, M. S. (2011). Inhibition of Sonic hedgehog and Notch pathways enhances sensitivity of CD133+ glioma stem cells to temozolomide therapy. *Molecular Medicine, 17*(1–2), 103–112. doi:10.2119/molmed.2010.00062.
32. Tegenge, M. A., Rockel, T. D., Fritsche, E., & Bicker, G. (2011). Nitric oxide stimulates human neural progenitor cell migration via cGMP-mediated signal transduction. *Cellular and Molecular Life Sciences, 68*(12), 2089–2099. doi:10.1007/s00018-010-0554-9.
33. Blurton-jones, M., Spencer, B., Michael, S., Castello, N. A., Agazaryan, A. A., Davis, J. L., et al. (2014). Neural stem cells genetically-modified to express neprilysin reduce pathology in Alzheimer transgenic models Neural stem cells genetically-modified to express neprilysin reduce pathology in Alzheimer transgenic models. *Stem Cell Research & Therapy, 5*(2), 46.
34. Farrel, K., Joshi, P., Roth, A., Kothapalli, C. R., & Lee, M.-Y. (2016). High-throughput screening of toxic chemicals against neural stem cells. In J. L. Sherley (Ed.), *Human stem cell toxicology* (pp. 31–63). Cambridge: The Royal Society of Chemistry.
35. Ferguson, C. S., & Tyndale, R. F. (2011). Cytochrome P450 enzymes in the brain: Emerging evidence of biological significance. *Trends in Pharmacological Sciences, 32*(12), 708–714. doi:10.1016/j.tips.2011.08.005.
36. Miksys, S., & Tyndale, R. F. (2013). Cytochrome P450-mediated drug metabolism in the brain. *Journal of Psychiatry & Neuroscience, 38*(3), 152–163. doi:10.1503/jpn.120133.
37. Murphy, S. V., & Atala, A. (2014). 3D bioprinting of tissues and organs. *Nature Biotechnology, 32*(8), 773–785. doi:10.1038/nbt.2958.
38. Arias, I., Wolkoff, A., Boyer, J., Shafritz, D., Fausto, N., Alter, H., & Cohen, D. (2011). *The liver: Biology and pathobiology*. Chichester: Wiley.
39. Boron, W. F., & Boulpaep, E. L. (2009). *Medical physiology* (2nd ed.). Philadelphia, PA: Saunders Elsevier.
40. Jaeschke, H. (2011). Reactive oxygen and mechanisms of inflammatory liver injury: Present concepts. *Journal of Gastroenterology and Hepatology, 26*(Suppl 1), 173–179. doi:10.1111/j.1440-1746.2010.06592.x.
41. Zimmermann, H. W., Trautwein, C., & Tacke, F. (2012). Functional role of monocytes and macrophages for the inflammatory response in acute liver injury. *Frontiers in Physiology, 3*, 1–18. doi:10.3389/fphys.2012.00056.
42. Kim, Y., & Rajagopalan, P. (2010). 3D hepatic cultures simultaneously maintain primary hepatocyte and liver sinusoidal endothelial cell phenotypes. *PloS One, 5*(11), 1–10. doi:10.1371/journal.pone.0015456.
43. Sato, Y., Tsukada, K., & Hatakeyama, K. (1999). Role of shear stress and immune responses in liver regeneration after a partial hepatectomy. *Surgery Today, 29*(1), 1–9.
44. Tuleuova, N., Lee, J. Y., Lee, J., Ramanculov, E., Zern, M. A., & Revzin, A. (2010). Using growth factor arrays and micropatterned co-cultures to induce hepatic differentiation of embryonic stem cells. *Biomaterials, 31*(35), 9221–9231. doi:10.1016/j.biomaterials.2010.08.050.
45. Ueno, T., Sata, M., Sakata, R., Torimura, T., Sakamoto, M., Sugawara, H., et al. (1997). Hepatic stellate cells and intralobular innervation in human liver cirrhosis. *Human Pathology, 28*(8), 953–959.
46. Messner, S., Agarkova, I., Moritz, W., & Kelm, J. M. (2013). Multi-cell type human liver microtissues for hepatotoxicity testing. *Archives of Toxicology, 87*(1), 209–213. doi:10.1007/s00204-012-0968-2.
47. Kostadinova, R., Boess, F., Applegate, D., Suter, L., Weiser, T., Singer, T., et al. (2013). A long-term three dimensional liver co-culture system for improved prediction of clinically relevant drug-induced hepatotoxicity. *Toxicology and Applied Pharmacology, 268*(1), 1–16. doi:10.1016/j.taap.2013.01.012.
48. Fiegel, H. C., Kneser, U., Kluth, D., & Rolle, U. (2010). Hepatic tissue engineering. *Handchirurgie, Mikrochirurgie, Plast Chir Organ der Deutschsprachigen Arbeitsgemeinschaft für Handchirurgie Organ der Deutschsprachigen Arbeitsgemeinschaft für Mikrochirurgie der Peripher Nerven und Gefässe Organ der Vereinigung der Deut, 42*(6), 337–41. doi:10.1055/s-0030-1252045.

49. Hoekstra, R., Nibourg, G. A., van der Hoeven, T. V., Plomer, G., Seppen, J., Ackermans, M. T., et al. (2013). Phase 1 and phase 2 drug metabolism and bile acid production of HepaRG cells in a bioartificial liver in absence of dimethyl sulfoxide. *Drug Metabolism and Disposition, 41*(3), 562–567. doi:10.1124/dmd.112.049098.
50. Bhatia, S. N., & Ingber, D. E. (2014). Microfluidic organs-on-chips. *Nature Biotechnology, 32*(8), 760–772. doi:10.1038/nbt.2989.
51. Toh, Y.-C., Lim, T. C., Tai, D., Xiao, G., van Noort, D., & Yu, H. (2009). A microfluidic 3D hepatocyte chip for drug toxicity testing. *Lab on a Chip, 9*(14), 2026–2035. doi:10.1039/b900912d.
52. Bale, S. S., Vernetti, L., Senutovitch, N., Jindal, R., Hegde, M., Gough, A., et al. (2014). In vitro platforms for evaluating liver toxicity. *Experimental Biology and Medicine (Maywood, N.J.), 239*(9), 1180–1191. doi:10.1177/1535370214531872.
53. Nakazawa, K., & Shinmura, Y. (2011). Effects of culture conditions on a micropatterned co-culture of rat hepatocytes with 3T3 cells. *Journal of Bioprocessing & Biotechniques, 01*(3). doi:10.4172/2155-9821.S3-002.
54. Li, C. Y., Stevens, K. R., Schwartz, R. E., Alejandro, B. S., Huang, J. H., & Bhatia, S. N. (2014). Micropatterned cell-cell interactions enable functional encapsulation of primary hepatocytes in hydrogel microtissues. *Tissue Engineering. Part A, 20*(617), 2200–2212. doi:10.1089/ten.TEA.2013.0667.
55. Ma, Y., Ji, Y., Huang, G., Ling, K., Zhang, X., & Bioprinting, X. F. (2015). 3D cell-laden hydrogel microarray for screening human periodontal ligament stem cell response to extracellular matrix. *Biofabrication, 7*(4), 044105. doi:10.1088/1758-5090/7/4/044105.
56. Bailey, S. N., Sabatini, D. M., & Stockwell, B. R. (2004). Microarrays of small molecules embedded in biodegradable polymers for use in mammalian cell-based screens. *Proceedings of the National Academy of Sciences of the United States of America, 101*(46), 16144–16149.

MIX
Papier aus verantwortungsvollen Quellen
Paper from responsible sources
FSC® C105338

If you have any concerns about our products,
you can contact us on
ProductSafety@springernature.com

In case Publisher is established outside the EU,
the EU authorized representative is:
Springer Nature Customer Service Center GmbH
Europaplatz 3, 69115 Heidelberg, Germany

Printed by Libri Plureos GmbH
in Hamburg, Germany